PHOTODETECTORS

An Introduction to Current Technology

UPDATES IN APPLIED PHYSICS AND ELECTRICAL TECHNOLOGY

Series Editor: P. J. Dobson
Philips Research Laboratories
Redhill, England

PHOTODETECTORS: **An Introduction to Current Technology**
P. N. J. Dennis

ERRATUM

Please note that pages 98 and 99
have been inadvertently interchanged

PHOTODETECTORS, P.N.J. Dennis   0-306-42217-4

A Continuation Order Plan is available for this series. A continuation order will bring delivery of each new volume immediately upon publication. Volumes are billed only upon actual shipment. For further information please contact the publisher.

PHOTODETECTORS
An Introduction to Current Technology

P. N. J. Dennis
Royal Signals and Radar Establishment
Great Malvern, England

Plenum Press • New York and London

Library of Congress Cataloging in Publication Data

Dennis, P. N. J.
  Photodetectors.

  (Updates in applied physics and electrical technology)
  Bibliography: p.
  Includes index.
  1. Photoelectronic devices. I. Title. II. Series.
TK8304.D46   1985   681′.415   85-30066
ISBN 0-306-42217-4

© 1986 Plenum Press, New York
A Division of Plenum Publishing Corporation
233 Spring Street, New York, N.Y. 10013

Text, excluding abstracts,
© 1986 Her Majesty's Stationery Office

Printed in the United States of America

PREFACE

This book has been written as part of a new series of scientific text-books being published by Plenum Publishing Company Limited. The scope of the series is to review a chosen topic in each volume, and in addition, to present abstracts of the most important references cited in the text. Thus allowing the reader to supplement the information contained within this book without have to refer to many additional publications.

This volume is devoted to the subject of Radiation Detectors, known as Photodetectors, and particular emphasis has been placed on devices operating in the infrared region of the electromagnetic spectrum. Although some detectors which are sensitive at ultraviolet and visible wavelengths, are also described. The existence of the infrared region of the spectrum has been known for almost two hundred years but the development of detectors specifically for these wavelengths was limited for a long time due to technology limitations and difficulties in understanding and explaining the phenomena involved. Significant advances were made during World War II, when the potential military applications of being able "to see in the dark" were demonstrated, and this progress has been maintained during the last forty years when many major advances have been achieved, such that the use of photodetectors for both civil and military applications is now relatively common and can be inexpensive.

The two most important methods of detecting this radiation are either by using the incident photon flux to excite carriers within a material or by using a substance with a strongly temperature dependent property. These different types of devices are reviewed and their properties discussed. Finally, a comparison of typical peerformance parameters are presented.

This book has been written such that no previous knowledge of the subject is required, and is intended as a review of the detection processes and presents details of current state-of-the-art photodetectors.

ACKNOWLEDGEMENTS

I would like to thank my many colleagues at RSRE for all their assistance during the preparation of this book. I also wish to thank Mullard Ltd, UK for their assistance in providing the following illustrations and photographs: Figures 4.10; 4.14; 4.17; 4.19; 4.20; 5.9; 5.12 and 5.18, GEC-Avionics Ltd and Rank Taylor Hobson UK for their assistance in providing the photographs used in Figure 5.13, and permission to use the following figures which are British Crown Copyright Reserved: Figures 3.6; 3.7; 4.18; 5.13 and 5.19.

I am also grateful to Miss J Bursnell for typing this volume and for all her patience on the numerous occasions when I have revised the text.

P N J Dennis

CONTENTS

CHAPTER 6

LIST OF FIGURES

CHAPTER 1

INTRODUCTION

The electromagnetic spectrum extends from the very short wave-lengths of X-rays ($\sim$ 0.1Å) to radiowaves, which can be thousands of metres in wavelength. The region of the spectrum encompassing the ultraviolet, visible and infrared radiation is termed the optical region and extends from 0.01 μm to 1000 μm. Radiation detectors suitable to operate in this region are called photodetectors.

The majority of these devices can be classified into two main categories: thermal detectors and photon detectors. A thermal detector is fabricated from a material with a strongly temperature dependent property which will be modified by the incident radiation absorbed and by measuring the change the amount of energy absorbed determined, for example a change in electrical conductivity or the expansion of a gas. The radiation is generally absorbed in a black surface coating giving a broad and uniform spectral response.

However, the sensitivity is generally lower than that obtained with a photon detector in which the incident radiation causes the excitation of electrons within the material and some form of electrical output signal is observed. The method of observing this output signal identifies two distinct categories of photon detectors, the photoemissive device and the solid state sensors. In the former the excited electrons are emitted from the photosensitive electrode into the surrounding medium, which may be either a vacuum or a gas. Whereas for solid state detector the photons cause a change in the electronic energy distribution within the material. This change is used in some way to produce an output signal. In a photoconductor, for example, the number of free electrons or holes is varied and the associated change in conductivity measured.

One parameter which is of great importance to all photon detectors is the quantum efficiency η, which is the ratio of the number of electron events per second occurring as a result of the irradiation, to the number of photons per second arriving at the detector. For an ideal device the quantum efficiency would be unity; however, even in high performance detectors it is generally limited to between 0.6 and 0.7.

The principle of operation of these various devices will be described in this book and their relative merits highlighted. It is intended to limit the discussion mainly to devices suitable for detection of infrared radiation in the wavelength region 0.7 to 1000 μm as this region has seen the most significant advances in detector technologies and applications during the last thirty years. However, it will be seen that many of the devices can be operated over a broad band which includes the infrared region.

One of the most significant features of the near infrared spectrum is that it contains the bulk of the thermal energy emitted from objects from ambient temperatures up to 6000K, the temperature of the sun. This is illustrated in Figure 2.1, which shows the spectral radiant emittance from an ideal blackbody source at different temperatures, as a function of wavelength.

The existence of the infrared region of the spectrum was first discovered by Sir William Herschel (1800), whilst investigating the energy distribution of the solar spectrum. He used a simple mercury-in-glass thermometer, which was traversed through the visible spectrum and he was surprised to discover that the energy spectrum peaked beyond the red end of the visible region. He repeated the experiment using alternative heat sources, a candle, a hot poker and a fire, and concluded that they all exhibited similar behaviour and the nature of the radiation was similar to the visible, differing only in momentum.

In the following year the existence of radiation at wavelengths shorter than the visible was discovered by Johann Wilhelm Ritter (1803). However, the experiment was severely limited by the primitive equipment available at the time. A major advance came in 1830 following the discovery of the thermoelectric effect by Seebeck, when Nobili (1830) and Melloni (1833) connected a number of thermocouples in series to increase the sensitivity and were able to study the emission spectrum from various sources and the transmission through many materials, including a rock salt prism.

The nature of infrared radiation was disputed for many years following the initial publication of these discoveries, and it was not until Fizeau and Foucault (1847) observed that interference fringes could be formed in exactly the same manner as with visible light, that the nature of the radiation was finally resolved. During the

following years much research was aimed at discovering the long
wavelength limit of the spectrum, and by filtering out of the short
wavelength radiation and using an interferometer the existence of the
spectrum to beyond 150 μm was proven. Gradually with the development
of Maxwell's electromagnetic radiation laws and the Stefan-Boltzmann
theory for blackbody radiation the behaviour and modelling of the
infrared spectrum became clear. These expressions are discussed more
fully in Chapter 2, with the figures of merit required to describe the
performance of various photodetectors, and in the following sections
the various detector types and their characteristics are discussed.

CHAPTER 2

DETECTOR PERFORMANCE

2.1 <u>Introduction</u>

The purpose of all photodetectors is to detect radiant energy of either a specific wavelength, or over a broad band, and to produce an output signal which is proportional to the amount of energy absorbed. Hence, in order to specify and compare the performance of various detectors it is necessary to define certain figures of merit which describe this conversion efficiency and the magnitude of the signal-to-noise ratio from the detector in terms of the incident radiation power. Other important parameters are the spectral response and the speed of response of the device to a changing input signal.

Let us first consider the properties of the emission of radiation from bodies at various temperatures. To describe the distribution of energy from any thermal radiator, the term blackbody was introduced by Kirchoff. This describes an object which absorbs all the incident radiation, independent of wavelength and conversely, is a perfect radiator. It was concluded by both Stefan in 1879 and Boltzmann in 1874 that the total amount of energy radiated from a blackbody is proportional to the fourth power of its absolute temperature. However, an expression for the spectral distribution of radiation from a blackbody source was not available until 1900 when Planck's equation was found to be in excellent agreement with experimental data. The Planck equation for the spectral radiant emittance W_λ, expressed in watts cm^{-2} µm^{-1} at a wavelength λ(µm) from a blackbody at absolute temperature T(K) is given in equation (2.1), where h is Planck's constant, in watts sec^2, c is the velocity of light and k Boltzmann's constant, in watts sec K^{-1}.

5

$$W_\lambda = \frac{2\pi hc^2}{\lambda^5} \left(\frac{1}{\exp\left(\frac{ch}{\lambda kT}\right) - 1} \right)$$ (2.1)

The blackbody radiation curves derived from this equation for objects at different temperatures are shown in figure 2.1, as a function of wavelength. The peak emission wavelength λ_m, of these curves is obtained from Wien's displacement law

$$\lambda_m T = 2898 \ \mu m \ K$$ (2.2)

which can be derived from Planck's equation.

It can be seen that the radiation is emitted as a continuous spectrum over a range determined by the temperature of the body. At low temperatures the emission is relatively low and mainly in the infrared, as the temperature is increased the radiant emittance increases rapidly, in proportion to the fourth power of the absolute temperature, and its peak shifts to shorter wavelengths, until at temperatures of a few thousand degrees Kelvin the peak of the curve occurs in the visible region of the spectrum.

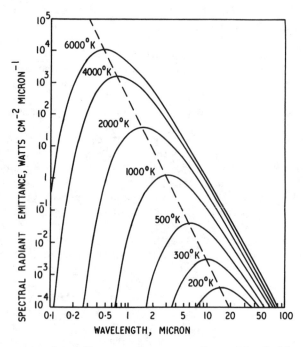

Fig. 2.1 The spectral radiant emittance of an ideal blackbody at various temperatures, as a function of wavelength in microns.

If Planck's law is integrated over all wavelengths from zero to infinity, the total radiant emittance W, is obtained, such that

$$W = \frac{2\pi^5 k^4}{15c^2h^3} T^4 \qquad (2.3)$$

$$W = \sigma T^4 \qquad (2.4)$$

this equation is known as the Stefan-Boltzmann law, and σ as the Stefan-Boltzmann constant.

When considering the performance of photon detectors it is often

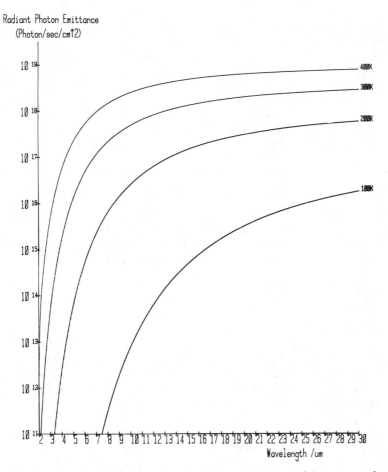

Fig. 2.2 The total radiant photon emittance incident on a detector up to its cut-off wavelength, as a function of the ambient temperature.

necessary to evaluate the photon flux emanating from a scene. The
spectral equations can be converted to photon radiance expressions
simply by dividing by the energy associated with one photon hc/λ, thus
the spectral radiant photon emittance Q_λ, in photons sec^{-1} cm^{-2} μm^{-1},
is given by

$$Q_\lambda = \frac{2\pi c}{\lambda^4}\left(\frac{1}{\exp(\frac{ch}{\lambda kT} - 1)}\right) \tag{2.5}$$

By integrating this expression the total photon emittance can be
derived, or if the integration is from zero to the detector wavelength
cut-off, the total flux incident on a photon detector up to its cut-
off will be derived, this is shown in Figure 2.2.

In most photodetection systems the radiation emitted by an object
must be transmitted through the atmosphere to the detector, which will
modify the emitted spectrum, in particular for applications over
significant ranges such as those over which most thermal imaging
systems are used. The atmosphere will absorb, scatter and refract
the radiation as a function of wavelength. The most important
absorption peaks are due to the molecular constituents of water
vapour, carbon dioxide, nitrous oxide and ozone; any small particles
suspended in the atmosphere will cause scattering of the radiation out
of the beam and also unwanted radiation into the optical path.
Finally, any turbulence in the atmosphere will cause refraction of the
signal.

The transmission of the atmosphere measured over a 6,000 feet
horizontal path at sea level is shown in Figure 2.3 (Hudson 1969), and
the most important absorbing molecules at each wavelength are
identified. The windows in the atmosphere between 3 and 5 μm and 8
and 14 μm, should be noted and as the peak of emission for objects
close to ambient temperature occurs close to 10 μm, the long
wavelength window is of major importance for many applications. The
3-5 micron window is also important as a better range of detectors is
available at these wavelengths.

In general, fog and cloud conditions produce strong scatterering
at all wavelengths, and consequently infrared systems in both bands
will be strongly affected by these adverse weather conditons, but the
transmission through rain is generally good.

Having considered the emission of radiation from a body and its
transmission through the atmosphere, the most important detector
parameters used to specify device performance will now be discussed,
and finally expressions for the ultimate performance of various types
of detectors will be derived.

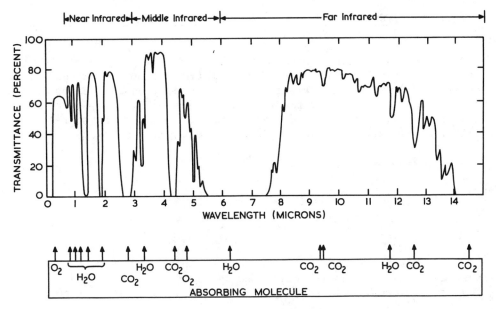

Fig. 2.3 The transmittance of the atmosphere for a 6000 ft.
         horizontal path at sea level containing 17 mm of
         precipitable water (Gebbie et al 1951).

2.2 Responsivity

One of the most fundamental properties of any detector is its
responsivity R, which is defined as the output signal per unit
input. If the radiation is chopped at a uniform frequency, the root
mean square voltage signal and radiant incident power are used in the
calculation. Thus the responsivity is given by

$$R = \frac{V_s}{P} = \frac{V_s}{HA} \qquad\qquad (2.6)$$

where V_s is the detector output voltage, for an incident power of P
watts, H is the value of the irradiance on the device in Wcm^{-2} and A
the sensitive area. It is necessary to specify the input radiation
source, which is generally a blackbody at 500 K or monochromatic
radiation at a wavelength λ.

Responsivity is generally quoted in either volts watt^{-1} or amps
watt^{-1} for infrared detectors but for visible detectors, such as
photomultipliers the responsivity is often given in microamps per
lumen, as the lumen is the standard unit of visible radiant power.

2.3 <u>Noise</u> <u>Equivalent</u> <u>Power</u> <u>(NEP)</u> <u>and</u> <u>Noise</u> <u>Equivalent</u> <u>Irradiance</u>
    <u>(NEI)</u>

Although the responsivity effectively defines the sensitivity of a
device it gives no indication of the minimum radiant flux that can be
detected. This minimum detectable flux is defined as the rms incident
radiant power required to produce an output signal V_s equal to the
detector noise level V_n, in other words, a signal-to-noise ratio of
unity, and is known as the noise equivalent power (NEP).

$$\therefore \quad NEP \quad = \quad \frac{P}{V_s/V_n} \qquad\qquad (2.7)$$

$$NEP \quad = \quad \frac{V_n}{R} \qquad\qquad (2.8)$$

When describing the performance of complete systems, the noise
equivalent irradiance is introduced as the radiant flux density
required to produce an output signal equal to the detector noise.

Thus

$$NEI \quad = \quad \frac{NEP}{A_L} \quad = \quad \frac{H}{V_s/V_n} \quad = \quad \frac{HV_n}{V_s} \qquad\qquad (2.9)$$

where A_L is the area of the collecting lens.

In a complete system the NEI is the flux density required at the
entrance aperture to give a signal-to-noise ratio of unity at the
output of the system electronics.

2.4 <u>Detectivity</u>

It can be seen that the higher the performance of a detector the
lower the value of NEP or NEI, to overcome this anomaly Jones (1953)
defined the detectivity D, of a device as the reciprocal of the NEP.
Thus,

$$D \quad = \quad \frac{1}{NEP} \qquad W^{-1} \qquad\qquad (2.10)$$

When the detectivity is used to characterise a detector it is
necessary to specify the wavelength of the incident radiation, the
detector temperature, the chopping frequency, any bias current applied
to the device, the area of the detector and the bandwidth of the
amplifier used to measure the detector noise.

The radiation source is often a blackbody at a standard

temperature, usually 500K, and the detector is generally operated at
an easily achievable temperature, for example, ambient (300K),
solidified carbon dioxide (195K) or liquid nitrogen (77K). The bias
current is normally optimised for maximum responsivity and the
chopping frequency set at a value sufficiently low that the detectivity is not limited by the time constant of the detector.

2.5 Specific Detectivity D*

The detectivity is not an ideal parameter for comparing different
detectors as it varies inversely as the square root of both the
sensitive area and the bandwidth Δf. Hence the specific detectivity
or D* (D-star) measured in $cm\ Hz^{\frac{1}{2}}\ W^{-1}$ has been introduced such that

$$D* = D(A\Delta f)^{\frac{1}{2}} \qquad\qquad (2.11)$$

$$D* = \frac{(A\Delta f)^{\frac{1}{2}}}{NEP} \qquad\qquad (2.12)$$

which is the signal-to-noise ratio when one watt of power is incident
on a detector having a sensitive area of $1\ cm^2$, and the noise is
measured over a 1 Hz bandwidth.

This is a very useful parameter for comparing detectors of
different sizes as the device noise is generally proportional to the
square root of the area. However, for a few detectors, including
Golay cells, many types of bolometers and metal-oxide-metal junctions
the detectivity is not inversely proportional to $A^{\frac{1}{2}}$, thus the
performance of these devices is normally reported as an NEP or
detectivity D.

The D* can be defined as a response to either a monochromatic or
a blackbody radiation source, and in the former case it is known as
the spectral D*, and presented in the form $D*(\lambda,f,\Delta f)$, where λ is the
wavelength at which the D* has been determined, which is usually the
peak response for a photon detector, f and Δf are the chopping
frequency and the bandwidth of the system, normally 1 Hz. Similarly
the blackbody D* is symbolised by $D*(T,f,\Delta f)$, where T is the
temperature of the blackbody, usually 500K.

2.6 Response Time and Frequency Response

The response time τ, for any detector is of great importance as
this will often determine the suitability of a device for a specific
application. This parameter is characterised by the speed of
response to a sudden change in the input signal.

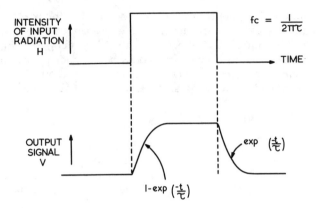

Fig 2.4 The detector response to a square wave input, defining the
detector response time τ.

The response time is specified as the time taken for the photosignal to rise to $(1 - 1/e)$ of the peak value or to decay to $1/e$ of
the peak value, as shown in Figure 2.4 when the detector is being
irradiated with pulsed radiation in which the rise and fall times are
so short that it appears as a square wave.

The frequency response is a plot of the output signal as a
function of the modulating frequency of the input radiation. In
general, this response V_s, is flat from zero up to a frequency,
usually in the megahertz range, at which it rolls off at 6 dB per
octave, such that

$$V_s = \frac{V_{so}}{(1 + (2\pi f\tau)^2)^{\frac{1}{2}}} \qquad (2.13)$$

where V_{so} is the zero frequency signal.

2.7 Spectral Response

When considering a detector for a particular use, as well as
knowing its sensitivity etc, it is essential to know the spectral
response, which describes the change in output signal as a function of
changes in the wavelength of the input signal. The response is
generally plotted as a function of wavelength for a constant radiant
flux per unit wavelength interval.

The response of a thermal detector is proportional to the energy
absorbed, thus they ideally exhibit a flat spectral response.
However, photon detectors have a response which is proportional to the
rate of arrival of photons, and as the energy per photon is inversely

proportional to wavelength, the spectral response increases linearly
with increasing wavelength, until the cut-off wavelength is reached,
which is determined by the detector material.

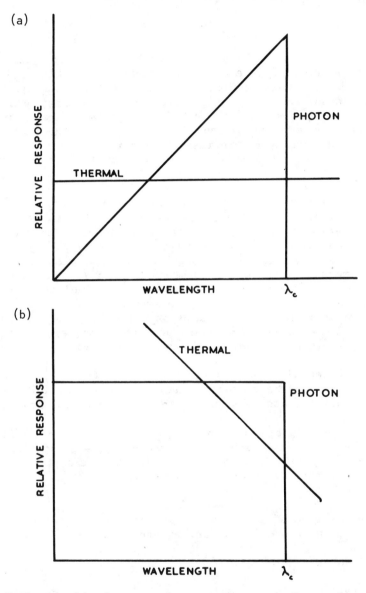

Fig. 2.5 The idealized spectral response curves for a photon and
thermal detector (a) for a constant flux per unit wavelength
interval, and (b) for a constant number of photons per unit
wavelength interval.

Figure 2.5(a) shows the spectral response of an ideal thermal and photon detector. Alternatively the spectral response is sometimes plotted for a constant rate of arrival of photons per unit wavelength interval and the shape of these curves are shown in Figure 2.5(b) with the photon detector having a flat response and the thermal device output decreasing with increasing wavelength.

2.8 Detector Noise Sources

The performance of all detectors will be limited by random noise processes, which will determine the minimum detectable power that can be measured. The noise sources may arise in the detector, in the incident radiant energy or in the electronic circuitry associated with the detection system. The object of any system design must be to reduce the electronic noise sources below that in the output of the detector, and ideally to minimise the internal detector noise such that the overall system noise is dominated by the incident radiation noise.

In this section only the internal noise mechanisms associated with the devices presented in this review will be discussed. More detailed treatments of noise theory can be found in many reviews, including Van der Ziel (1959) and Van Vliet (1967).

Noise mechanisms result from stochastic or random processes, consequently it is not possible to predict the amplitude at any particular time. The value of the noise is normally specified in terms of the root mean square value of the voltage or current, and if two uncorrelated noise sources are present in any system their net effect is determined by adding the mean square values.

The fundamental noise process in the absence of an electrical bias arises due to the random motion of the free carriers within any resistive material and is known as the Johnson or thermal noise.

In a photoconductive element a second noise source is due to the carrier density fluctuations and is known as the generation-recombination (g-r) noise. For a junction device shot noise of the diffusing carriers is observed, and thermal detectors exhibit temperature noise.

A final source of excess noise is known as 1/f noise, or flicker noise as it exhibits a 1/f power law spectrum.

2.8.1 Johnson Noise

This form of noise is found in all resistive materials, and occurs in the absence of any electrical bias. It was first demonstrated in 1928, by Johnson, and the open circuit noise voltage V_n and short circuit noise current i_n, are given in equations (2.14) and (2.15).

$$V_n = (4kTr\Delta f)^{\frac{1}{2}} \tag{2.14}$$

$$i_n = \left(\frac{4kT\Delta f}{r}\right)^{\frac{1}{2}} \tag{2.15}$$

where k is Boltzmann's constant, T the absolute temperature of the element, r its resistance, and Δf the electrical bandwidth.

It can be seen that this noise source is independent of frequency, that is white noise, and for a given sample just depends on its temperature.

2.8.2 Generation-Recombination (G-R) Noise

Generation-recombination noise arises from the fluctuations in the rates of optical and thermal generation and recombination of free carriers in a semiconductor, thus causing a variation in the average carrier concentration, and hence the electrical resistance. The spectrum for a g-r noise source is flat upto a value that is approximately equal to the inverse of the free carrier lifetime, it then rolls off at roughly 6 dB per octave.

Expressions for the g-r noise have been derived by Long (1967)[1] and Kruse (1977), two of the most common examples are for a simple one-level extrinsic semiconductor and for a near intrinsic semiconductor.

For extrinsic devices the short circuit g-r current is given by

$$i_n = 2I_B \left(\frac{\tau \Delta f}{N_o(1 + \omega^2\tau^2)}\right)^{\frac{1}{2}} \tag{2.16}$$

and the open circuit voltage

$$V_n = 2I_B r \left(\frac{\tau \Delta f}{N_o(1 + \omega^2\tau^2)}\right)^{\frac{1}{2}} \tag{2.17}$$

where I_B is the bias current, r the sample resistance, τ the free carrier lifetime, N_0 the total number of free carriers and ω the angular frequency.

In the case of the near instrinsic detector the fluctuations arise due to variations in the intrinsic generation, and if the sample is n type and the electron mobility greater than that of the holes, the g-r short circuit current and open circuit voltage approximate to

$$i_n = \frac{2I_B}{N} \left(\frac{p \tau \Delta f}{1 + \omega^2 \tau^2} \right)^{\frac{1}{2}}$$

(2.18)

$$V_n = \frac{2I_B r}{N} \left(\frac{p \tau \Delta f}{1 + \omega^2 \tau^2} \right)^{\frac{1}{2}}$$

(2.19)

where N and P are the total number of free electrons and holes in the sample.

2.8.3 Shot Noise

Shot noise is the noise source found in junction devices and arises from the random thermal motion which causes variations in the diffusion rates in the neutral zone of the junction and generation-recombination fluctuations occurs in both the neutral and depletion regions.

A detailed, general theory of diode noise mechanisms has not been derived, although Buckingham and Faulkner (1974) have derived expressions for the ideal diffusion limited case.

The usual form of diode current–voltage characteristic is given by

$$I = I_o \left[\exp\left(\frac{qV}{kT} \right) - 1 \right]$$

(2.20)

where I is the diode current, I_o the reverse bias saturation current, V the applied voltage, q the electronic charge and T is the absolute temperature.

The shot noise current is then given by

$$i_n = [(2qI + 4qI_o) \Delta f]^{\frac{1}{2}}$$

(2.21)

and the zero bias resistance, r_o by

$$r_o = \frac{kT}{qI_o}$$

(2.22)

Thus at zero bias, I = 0, the shot noise is given by

$$i_n = \left(\frac{4kT\Delta f}{r_o}\right)^{\frac{1}{2}} \qquad (2.23)$$

which is equivalent to the expression for the Johnson noise current associated with the zero bias slope resistance. When operated under reverse bias $I = -I_o$, the noise current is given by

$$i_n = (2qI_o\Delta f)^{\frac{1}{2}} \qquad (2.24)$$

Photoemissive, vacuum tubes also exhibit shot noise, caused by variations in the rate of arrival of electrons at the collecting electrode. The expression for this form of shot noise is similar to that of equation (2.24), but now I_o is the dark current from the photocathode, arising from thermionic emission, field emission or anode-cathode leakage currents.

It can be seen that the power spectrum of shot noise will be flat.

2.8.4 Temperature Noise

One source of noise observed only in thermal detectors is temperature noise, which arises from the changes in the temperature of the detector which in turn is due to the fluctuations in the rate at which heat is transferred from the detector to its surroundings.

If the detector is represented as a thermal capacitance which is connected to a large heat sink at a constant temperature T, via a thermal conductance G, in the absence of any radiation input the average temperature of the detector will also be T, but the fluctuations in this value will give rise to a source of detector noise, known as the temperature noise, which will have an rms fluctuation in power given by (Smith et al 1968):

$$\Delta W_I = (4kT^2G)^{\frac{1}{2}} \qquad (2.25)$$

The power spectrum for temperature noise is also flat.

2.8.5 1/f Noise

One source of noise exhibited both in semiconductors and phototubes is known as 1/f noise, as the spectrum of the noise power approximates to an inverse frequency dependence.

There is no exact expression for the noise current, but it is of the form

$$i_n = \left(\frac{kI_B^\alpha \, \Delta f}{f^\beta}\right)^{\frac{1}{2}}$$
(2.26)

where K is a proportionality constant, I_B is the bias current and α and β are characteristics of the device. In most cases α is approximately equal to 2 and β is in the range 0.8 to 1.5.

In semiconductor devices this noise source is associated with potential barrier effects at the contacts, surface traps and surface leakage currents.

In order to prevent 1/f noise dominating detector performance careful device fabrication is required, and in the case of photodiodes, operation of the detector at zero bias reduces its contribution.

2.9 Summary

An idealised noise spectrum for a photoconductive element is shown in Figure 2.6, in the absence of any radiation. At low frequencies the 1/f noise dominates, at intermediate frequencies the g-r noise and at high frequencies the Johnson noise is the most important. The knee points of the spectrum vary from detector to detector but approximately occur in the 1 KHz and 1 MHz regions.

As previously discussed even if all of these noise sources in a detector could be eliminated, there would still be a fundamental limit to a detector's performance due to fluctuations in the background radiation. The fluctuations in this emission process give rise to a noise equivalent power and hence a maximum value of the detectivity which can be achieved for any detector, this is the background limited photodetector (BLIP) value.

For most situations the surroundings will be at approximately 300 K, which corresponds to a maximum intensity near to 10 μm wavelength.

In the case of an ideal thermal detector, operating at room temperature, it is assumed that the response of the device is independent of wavelength and that it receives background radiation emanating from a complete hemispherical field of view. In the absence of external noise sources such as the Johnson noise in any resistors or amplifier noise the only noise source will be the temperature noise (see section 2.8.4). Thus the mean square fluctuation in radiation power is given in equation (2.27).

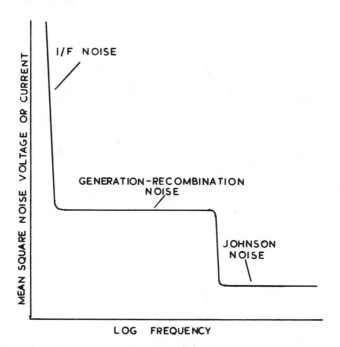

Fig. 2.6 The relative contributions of noise sources for a
photoconductive detector, in the absence of any radiation.

$$\Delta W_T = (4kT^2 G \Delta f)^{\frac{1}{2}} \qquad (2.27)$$

If the thermal conduction is purely by radiation, ΔW_T will be a be
a minimum and G is determined by

$$G = 4\sigma A \eta T^3 \qquad (2.28)$$

where σ is the Stefan–Boltzmann constant and η the emissivity. For
these ideal conditions the NEP can be evaluated such that

$$\eta(NEP) = \Delta W_T = (16\sigma \eta k T^5 A \Delta f)^{\frac{1}{2}} \qquad (2.29)$$

$$\therefore \qquad NEP = \left(\frac{16\sigma k T^5 A \Delta f}{\eta} \right)^{\frac{1}{2}} \qquad (2.30)$$

Thus for an ideal thermal detector with a unit sensitive area and
unit bandwidth and an emissivity of one

$$D* = (16\sigma kT^5)^{-\frac{1}{2}} \qquad (2.31)$$

For example if the detector is operating into a background maintained at 300 K, assumed to emit as a blackbody

$$D* = 1.81 \times 10^{10} \text{ cm Hz}^{\frac{1}{2}} \text{ W}^{-1} \qquad (2.32)$$

However, unlike the thermal devices, photon detectors do not have a flat spectral response, and they will only respond to photons of energy greater than the minimum energy E, required to excite the electronic transition. Hence, it is assumed that for $hv \gg E$, the quantum efficiency η is one; whilst if $hv < E$, $\eta = 0$.

Kruse et al (1962) have derived an expression for the NEP for a photovoltaic detector, assuming the background flux greatly exceeds the signal flux, thus

$$NEP = E_\lambda \left(\frac{2A \phi_B \Delta f}{\eta}\right)^{\frac{1}{2}} \qquad (2.33)$$

where E_λ is the photon energy of the signal radiation, and ϕ_B is the total background photon flux to which the detector is sensitive.

Hence the background limited value of the D* for a photovoltaic detector is given by

$$D* = \frac{\lambda}{hc} \left(\frac{\eta}{2\phi_B}\right)^{\frac{1}{2}} \qquad (2.34)$$

which for an ideal detector corresponds to

$$D* = 3.56 \times 10^{18} \lambda \left(\frac{1}{\phi_B}\right)^{\frac{1}{2}} \qquad (2.35)$$

In the case of a photoconductive element an equilibrium is established between the generation and recombination of carriers, and the fluctuations in these rates cause an additional noise source, which is absent for a photodiode. This gives an increase in the NEP of $\sqrt{2}$, such that

$$NEP_{(photoconductor)} = 2E_\lambda \left(\frac{A\phi_B \Delta f}{\eta}\right)^{\frac{1}{2}} \qquad (2.36)$$

$$D^*(\text{photoconductor}) \;=\; 2.52 \times 10^{18}\,\lambda\left(\frac{1}{\phi_B}\right)^{\frac{1}{2}} \qquad (2.37)$$

The photon flux ϕ_B, is the total flux from zero wavelength upto the detector cut-off wavelength. These fluxes are shown in Figure 2.2, as a function of the blackbody temperature, operating into a 2π field of view. Hence the peak detectivity of both photoconductive and photovoltaic photon detectors can be derived as a function of the cut-off wavelength, and this is shown in Figure 2.7 for a 300K background temperature. A minimum value of D* occurs at a cut-off wavelength of approximately 14 µm, if this cut-off value is reduced the D* increases rapidly as the background photon flux falls; at longer

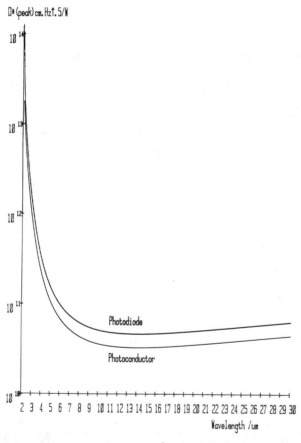

Fig. 2.7 The theoretical value of D^* (peak) for a background limited photoconductor and photovoltaic detector as a function of cut-off wavelength, for a 2π field of view at a temperature of 300K.

wavelengths the background flux increases, but not as rapidly as the signal flux, hence the D* increases towards longer wavelengths.

It has already been pointed out that the BLIP value of photoconductive and photovoltaic detectors differs by a factor $\sqrt{2}$, and it is also true that the value is a function of the background temperature, detector operating temperature and the field of view. These variations have been discussed in detail by Kruse (1977), but in summary the variation of D* as a function of the background temperature is shown in Figure 2.8. If the detector operating temperature is near ambient, it will receive significant radiation

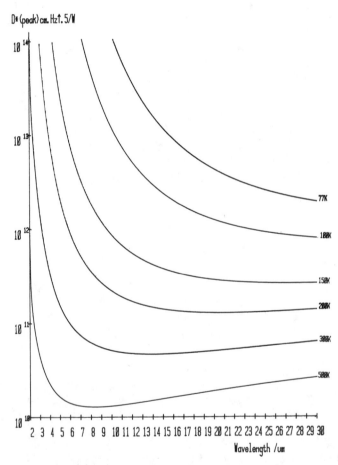

Fig. 2.8 The theoretical value of D* (peak) for a background limited photovoltaic detector as a function of the background temperature, from 77 K to 500 K, for a 2π field of view.

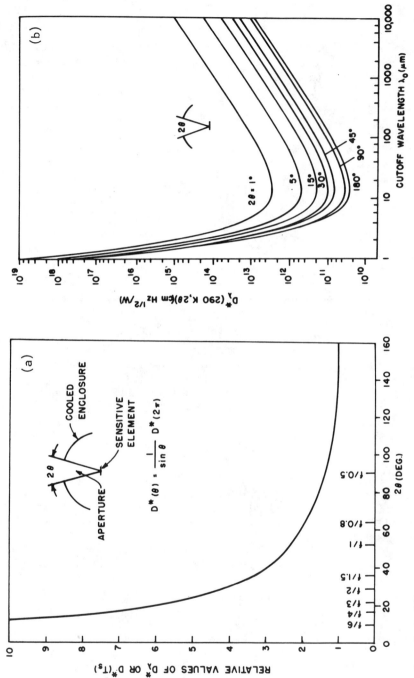

Fig. 2.9 (a) The relative increase in photon noise limited D* (peak) achieved by using a cooled aperture in front of the detector (Kruse et al 1962) (b) The improvements in D* (peak) as the field of view is reduced, for a background limited photoconductor, as a function of wavelength. For a 290 K background temperature.

both from the normal forward direction, and also from the rear.
Depending on the device structure the D* (BLIP) can be reduced by $\sqrt{2}$,
but for devices operating at well below room temperature this effect
can be ignored.

The most significant detector performance increases are achieved
by restricting the field of view. All the values so far discussed
have applied to a 2π (or hemispherical) field of view. If the
background photon flux is reduced the detectivity will increase for
background limited detectors. This can be achieved by use of either a
cooled aperture to limit the angular aperture of the detector or a
cooled filter to limit the spectral response. If the cold shield
restricts the field of view to 2θ, the background flux will be
reduced such that

$$\phi_B(2\theta) \quad = \quad \phi_B(2\pi) \; \sin^2\theta \tag{2.38}$$

and hence the D* will increase to

$$D_\lambda^*(BLIP)(2\theta) \quad = \quad \frac{D_\lambda^*(2\pi)}{\sin\theta} \tag{2.39}$$

The improvement in D* obtained by reducing the field of view is
shown in Figure 2.9(a) and(b). A new detector parameter, D** (dee
double star) has been introduced to remove the dependence of detector
performance on the field of view, and it is the value of D* referenced
to an effective weighted solid angle of π steradians, which is
approximately equivalent to a hemispherical surround, such that

$$D** \quad = \quad \left(\frac{\Omega}{\pi}\right)^{\frac{1}{2}} D* \tag{2.40}$$

where Ω is the effective weighted solid angle the detector sees
through the cold shield. If the device is Lambertian (that is a
perfectly diffusing surface), has a circular symmetry and if the solid
angle subtended by the aperture at the detector can be represented as
a cone with half angle θ, then

$$\Omega \quad = \quad \pi \sin^2\theta \tag{2.41}$$

$$D** \quad = \quad D*\sin\theta \tag{2.42}$$

The second method involves the use of a cooled optical filter
mounted in front of the detector, chosen such that the maximum amount
of background photon flux is rejected whilst rejecting a minimum of
the target flux.

CHAPTER 3

THERMAL DETECTORS

3.1 Introduction

By monitoring the change in temperature of an object it is possible to deduce the amount of radiant energy absorbed. Thermal detectors achieve this by utilizing a material with a strongly temperature dependent property such as the electrical conductivity or the thermal expansion. Herschel's original experiments to investigate the electromagnetic spectrum used a blackened liquid in glass thermometer which can be considered as the first manufactured thermal device. However, it has recently been discovered that thermal sensors have existed in nature for many thousands of years; two families of snakes, the pit vipers and boa constrictors, use such sensors, mounted in their head, to locate their prey (Bullock and Barrett 1968).

Due to the nature of thermal detectors they have a broad flat spectral response, compared to photon devices. However, one disadvantage is their relatively slow response times, typically of the order of milliseconds, which limits their use in certain applications. Some of the more common thermal detectors are listed in Table 3.1, with their temperature dependent properties.

The relatively insensitive liquid and gas thermometers used by Herschel were soon replaced by the first true infrared detector, the thermopile, designed by Nobili in 1829 (Nobili 1830) and used by Melloni (see Scott Barr 1962). This was followed some years later by a resistive metal bolometer invented by Langley. Since these early detectors many advances have been made in improving their sensitivity and reliability, and one of the most significant advances has been the pyroelectric device which has enabled relatively inexpensive detectors

25

Table 3.1   Thermal Detectors

Detectors	Method Of Operation
Thermopile	Voltage generation, caused by the change in temperature of the junction of two dissimilar materials
Bolometers	Change in electrical conductivity
Golay cell	Thermal expansion of a gas
Pyroelectric and pyro-magnetic devices	Changes in electrical and magnetic properties
Evapograph	The rate of vapour condensation on a thin membrane
Thermionic emission	Rate of emission from an oxide coated cathode
Absorption edge image converter	Optical transmission of a semi-conductor
Liquid crystal device	Changes of optical properties

to be produced in large quantities, and recently Roundy et al (1974) have produced pyroelectic devices with picosecond response times.

When considering the performance of a thermal detector it is necessary to determine the temperature rise of the system due to the incident radiation and secondly the resulting output signal due to the change in the detector's temperature dependent property. The first part of this calculation is common to all thermal devices but the second process will differ from device to device.

In Section 2.9 an expression for the NEP of a thermal detector was derived as

$$ NEP = \left(\frac{16\sigma \ kT^5 \ A \ \Delta f}{\eta} \right)^{\frac{1}{2}} \qquad (2.30) $$

Consequently the detectivity of an ideal device of unit sensitive area, unit bandwidth and an emissivity of one, detecting radiation from a 300K blackbody source is given by

$$D^* = 1.81 \times 10^{10} \text{ cm Hz}^{\frac{1}{2}} \text{ W}^{-1} \quad\quad\quad\quad\quad (2.32)$$

It can be seen that an improvement in D^* could be achieved by operating the detector in a cooled environment, which would be the case for a device used in outer space. At the Universe's background temperature of approximately 3K the detectivity would increase to $1.81 \times 10^{15} \text{ cm Hz}^{\frac{1}{2}} \text{ W}^{-1}$. Some detectors, bolometers in particular, have been operated at liquid helium temperatures to try and achieve this improved performance. However, this cannot be fully realised, until the complete experiment is carried out in an environment maintained at liquid helium temperatures, as radiation will always be admitted through the detector entrance aperture from a higher temperature background.

The principle of cooling detectors to improve their performance is mainly applicable to bolometers and pyroelectric devices, as with many other thermal sensors their temperature dependent properties reduce as they are cooled. Also, the requirement to cool these devices tends to detract from one of their principle advantages over many photon detectors.

Some of the more important thermal detectors will be described in the following sections and their relative performances and applications discussed.

3.2 Thermopile

One of the original infrared detectors was the thermopile, based on the Seebeck effect. He discovered in 1826 that at the junction of two dissimilar conductors a voltage could be generated by a change in temperature. Using this effect Melloni produced the first thermopile detector in 1833, to investigate the infrared spectrum.

To produce an efficient device the junction thermal capacity must be minimized, to give as short a response time as possible, and the absorption coefficient optimized, which is often achieved by blackening the sensor. The junction should be fabricated from two conductors with a large Seebeck coefficient θ, and low thermal conductivity k, to minimize the heat transfer between the hot and cold junctions, and a large electrical conductivity σ, to reduce the heat developed by the flow of current. Unfortunately these requirements are incompatible and the best compromise is obtained when $(\sigma\theta^2)/k$ is a maximum.

The first thermopiles were constructed from fine metallic wires, the most popular combinations being bismuth-silver, copper-constantan and bismuth-bismuth/tin alloy. The two wires are joined to form the thermoelectric junction, and a blackened receiver, usually a thin gold foil which defines the sensitive area, is attached directly to the junction.

The development of semiconductors produced materials with much larger Seebeck coefficients, and hence the possibility of constructing thermopiles with increased sensitivities. However, due to difficulties in producing fine wires from these materials and making contact between them, a new fabrication technique was required. Schwartz (1952), developed a method in which the two semiconductors are fused on to the tips of a pair of gold pins, and a gold foil is then melted between the pins, which is then blackened and acts as the receiver. The most common alloys used for this construction are (33% Te, 32% Ag, 27% Cu, 7% Se, 1% S), for the positive electrode and (50% Ag_2Se, 50% Ag_2S) for the negative.

The sensititity of these devices can be increased by about an order of magnitude if they are mounted in a vacuum or filled with a gas which has a low thermal conductivity, for example xenon, and detectivities of 3×10^9 cm $Hz^{\frac{1}{2}}$ W^{-1} have been achieved. However this has the effect of increasing the response time to typically 30 msec. The response time can be reduced by reducing the thickness of the deposited film, but the Johnson noise will increase due to the increase in resistance. The spectral reponse of these devices is determined by the transmission of the encapsulation window, and the efficiency of the gold foil to act as a perfect absorber.

By careful design it is possible for a thermopile to be 99% efficient from the visible to beyond 40 µm. The high performance of these room temperature devices, only recently bettered by pyroelectric detectors, makes their use ideal for many spectrometers (Fellgett 1972).

Although semiconductor thermopiles have higher sensitivities than the conventional metal wire devices, they are less robust and stable, and consequently if a high degree of reliability and long term stability is required metal thermopiles will be used, for example as industrial radiation pyrometers for high temperature measurements in steel plants and ground based meteorological instruments for measuring the radiant intensity of the sun (Drummond 1970).

To fabricate much faster thermopiles, vacuum deposition has been investigated to produce metal junctions, and with modern techniques small areas and complex arrays can be fabricated. These devices are cheap and much more rugged than the traditional devices, and consequently have been used successfully in a number of space instruments. Bismuth and antimony are the two most common materials used, they are evaporated onto an insulating substrate with a good thermal conductivity which forms a heat sink, such as sapphire or beryllium oxide, the hot junction is often coated with black carbon

soot, and the cold junction with a reflecting film. Devices
fabricated in this way can have response times of less than 30 nsec.
Unfortunately they have relatively low sensitivity, with
responsivities of 5×10^{-6} VW^{-1} and D* of 10^6 cm $Hz^{\frac{1}{2}}$ W^{-1}.
Consequently they are used as heterodyne detectors of CO_2 laser
radiation, where the signal power levels are high, but fast response
times are required (Contreras and Gaddy 1971).

However, the sensitivity of these devices can be improved
substantially by fabricating the junction on a thin polymer substrate,
which will make a negligible contribution to the thermal
characteristics (Stevens 1970), and responsivities of 10^2 VW^{-1} and
D* 3 $\times 10^8$ cm $Hz^{\frac{1}{2}}$ W^{-1} have been obtained, but the response times are
reduced significantly, to 1 msec.

With this technology these detectors have been fabricated in
various geometries and arrays containing over one hundred elements,
have been produced.

3.3   Bolometers

A simple bolometer consists of a resistor with a very small
thermal capacity, and a large temperature coefficient, and hence,
significant changes in conductance occur if the detector is heated by
incident radiation. The device is operated by passing an accurately
controlled bias current through the detector, and monitoring the
output voltage.

The first bolometer produced by Langley in 1880 (Langley 1881)
consisted of a blackened thin platinum foil, and was used for solar
observations. These metal bolometers, fabricated either from thin
foils or using evaporated layers such as nickel, bismuth or antimony,
are still in use today where long term stability is required. The
devices are operated at room temperature, and have specific
detectivities of the order 1×10^8 cm $Hz^{\frac{1}{2}}$ W^{-1}, with response times of
approximately 10 msec. Unfortunately they are generally rather
fragile, thus limiting their use in certain applications. However
this problem has now been overcome by the use of the thermistor
bolometer, which is constructed from a sintered mixture of various
semiconducting oxides which have a higher temperature coefficient of
resistance than metals, and are generally much more rugged. The
temperature coefficient depends on the band gap, the impurity states
and the dominant conduction mechanism. A mixture commonly used for
infrared detectors contains nickel, manganese and cobalt, prepared in
thin flakes, 10 μm thick, mounted on an electrically insulating
substrate, for example sapphire, which is then mounted on a metallic
heat sink, to control the time constant of the device. The sensitive
area is blackened to improve its radiation absorption characteristics.
The device sensitivity and response time cannot both be optimised as
improved heat sinking to reduce the time constant prevents the

detector from reaching its maximum temperature, thus reducing the responsivity. D* values of $1.6 \times 10^8 \times \tau^{\frac{1}{2}}$ cm $Hz^{\frac{1}{2}}$ W^{-1}, where τ is the time constant, have been obtained for these devices. The construction of this type of bolometer is shown in Figure 3.1.

They are generally operated in a similar manner to the metal bolometer, by using a pair of detectors mounted in opposite arms of a bridge circuit and using one device as a control, screened from the radiation source.

Thermistor bolometers were developed during World War II in the United States for infrared spectroscopy and heat sensing applications, and have since been used successfully in satellite instrumentation, burglar alarms and fire detection systems.

To reduce the cost of bolometers the possibility of using evaporated films has been investigated, chalcogenide glasses, such as $Tl_2SeAs_2Te_3$ (Bishop and Moore 1973) and sputtered amorphous films, Ge_xH_{1-x} (Moustakas and Connell 1976) have been prepared.

A significant improvement in the performance of bolometer detectors is achieved if the devices are cooled, as the resistance changes are much greater than at room temperature, and the device can be fabricated thicker to improve the infra red absortion without increasing the thermal capacity, due to the reduced specific heat in the cooled material. By surrounding the detector with a cooled enclosure the sensitivity can be increased by at least two orders of magnitude over the room temperature value, and the ultimate performance will now be determined by the background radiation entering through the front aperture (Putley 1964).

The first cooled bolometers were produced from superconductors such as tin, with a transition temperature in the helium temperature region (Martin and Bloor 1961). Unfortunately their usefulness was

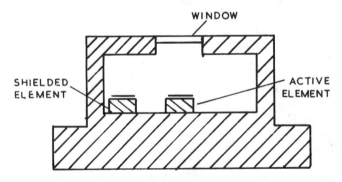

Fig. 3.1 The construction of a thermistor bolometer, showing the active and shielded element.

limited due to the stringent temperature control required, their poor radiation absorption characteristics due to being thin and their fragility.

The performance was generally limited by amplifier noise rather than radiation fluctuations, however NEP's of about 3×10^{-12}, with $\tau \sim$ 1.25 sec. were achieved. Recently, some robust devices with response times of the order of 3 µsec have been reported by Gallinaro and Varone (1975). They fabricated the detector from a tin film evaporated onto an aluminium substrate, maintained at 3.7K, the transition temperature for tin. NEP values of $\sim 10^{-13}$ $WHz^{-\frac{1}{2}}$ have been reported for this device, however the temperature must still be stabilized accurately to within 10^{-5}K.

An improvement on these superconducting bolometers are the cooled semiconductor devices which do not require such severe temperature stability and have higher absorption coefficient. Carbon was used in the first samples (Boyle and Rodgers, 1959), but improved performance was obtained with doped germanium or silicon elements (Low 1961, 1965, 1966). They are used extensively in infrared astronomy, where over most of the far infrared spectrum their performance is comparable to that of the best photon detectors with the added advantage of being a broad band device. For most of these applications the relatively long time constant is not a disadvantage (Coron 1976). Zwerdling (1968) described a bolometer constructed from a single crystal of germanium doped with gallium, to increase the long wavelength absorption. The device is operated in a pumped liquid helium cryostat at a temperature of 1.55K.

By using suitable impurity dopants germanium devices can exhibit extrinsic photoconductive properties to beyond 100 m. If the doping levels are increased, it is necessary to counter dope the material to maintain its high resistivity and temperature coefficient of resistance, but its response can be extended to the submillimeter region. The radiation absorbed is transferred rapidly to the crystal lattice, raising its temperature, rather than producing photoconduction. The absorption efficiency can be increased by mounting the device in an integrating cavity. These devices have been discussed by Putley (1977).

Recent designs of detectors have increased their performance significantly. Drew and Sievers (1969) have obtained an NEP of approximately 10^{-14} $WHz^{-\frac{1}{2}}$ for a device operating at 0.5K. If the temperature is reduced to 0.1K, values of less than 10^{-15} $WHz^{-\frac{1}{2}}$ and responsivities of 3×10^7 VW^{-1} can be achieved (Draine and Sievers 1976).

The use of silicon has been studied recently as an alternative to germanium, as it has a lower specific heat, easier materials preparation, and a more advanced device fabrication technology.

Silicon bolometers with NEP's of 2.5×10^{-14} WHz$^{-\frac{1}{2}}$ which compare favourably with germanium, have been reported by Kinch (1971). Nayar (1974) has described a system, using thallium selenide, and Von Ortenberg et al (1977), have suggested stannic oxide as a bolometric material for use where the detector may be subjected to applied magnetic fields for example in magnetospectroscopy.

When discussing the performance of bolometers the NEP is probably a better parameter for comparison than D*, as two of the principal noise sources, from the electrical leads and the amplifier, are independent of detector area.

In order to extend the wavelength spectrum to the short millimetre region 'composite bolometers' have been developed (Clarke et al 1977, Richards et al 1976). The radiation is absorbed by a bismuth film, deposited onto a sapphire substrate, or an epitaxial layer of silicon on sapphire and the heat is then transferred by conduction to the bolometer element mounted on the opposite side of the substrate which acts as a sensitive thermometer. The device is operated at the super-conducting transition temperature of 1.3K. A detector of this type is illustrated in Figure 3.2.

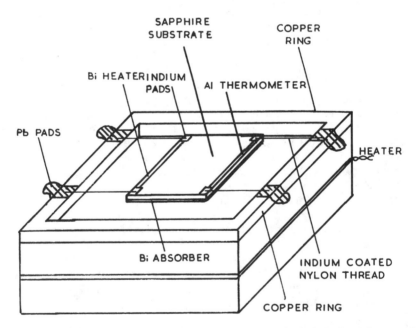

Fig. 3.2 A composite bolometer using a bismuth film absorber and an aluminium superconducting-transition thermometer (Clarke et al 1977).

With these devices an electrical NEP of 2×10^{-15} cm $Hz^{-\frac{1}{2}}$ has been
achieved. One of the fundamental limits on performance may be due to
the low noise cryogenic amplifiers required, however this is within a
factor two of the fundamental thermal noise limit. Hauser and Notary
(1975) have reported the use of this type of detector in an
astronomical telescope for millimeter wave continuum observations.

The sensitivity of these detectors is partly limited by the
thermal coupling to the substrate and the presence of liquid helium or
low pressure helium gas in the sample chamber. Chin (1977) has
recently reported results using a superconducting granular aluminium
film suspended between two contact leads, cooled by helium gas. The
device had an NEP of 4×10^{-17} $WHz^{-\frac{1}{2}}$ and a response time of 100 nsec.

3.4 Golay Cell

The concept of monitoring the pressure changes in a gas due to
radiation heating has been known for many years, and a constant
pressure gas thermometer was used by Leslie to study radiant heat.
However, with the development of the bolometer and thermopile the
concept was not pursued until 1947, when Golay developed a constant
volume cell which consists of an enclosed volume of gas, usually xenon
due to its low thermal conductivity, at a low pressure, and a thin
metal membrane which absorbs the incident radiation. Radiation
incident on the cell warms the gas and thus its pressure increases,
causing part of the cavity wall, which is constructed from a silvered
flexible membrane, to distort. This movement deflects a light beam
which shines onto a photocell, producing a change in ouput signal
level, as shown in Figure 3.3. The performance of the Golay cell is
only limited by the temperature noise associated with the thermal
exchange between the absorbing film and the detector gas, consequently
the detector can be extremely sensitive, with NEP's of approximately
2×10^{-10}, $D* \sim 3 \times 10^{9}$, and responsivities of 10^{5} to 10^{6} VW^{-1}.

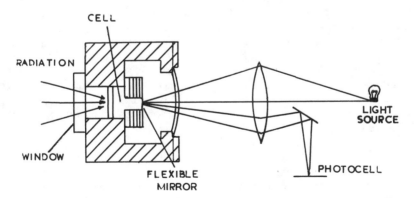

Fig. 3.3 The construction of a Golay cell using a light source and
         photocell.

However due to its construction the detector is fragile and very
susceptible to vibration and is only suitable for use in a controlled
environment such as a laboratory. Its response time is also quite long
typically 15 msec.

The modern Golay cells have overcome some of these problems by
replacing the photocell with a solid state photodiode and using a
light emitting diode for illumination (Hickey and Daniels 1969).
These components also reduce the heat dissipation inside the cell,
and increase its reliability.

An alternative scheme, used by Chatanier and Gauffre (1973), is to
use the distorted membrane and a conducting electrode placed close by
as a variable capacitor, which is monitored. This cell is
sufficiently ruggedised to allow its adoption for space applications.
It is also used extensively as a gas analyzer (Luft cell) in which
the gas being investigated is placed in one cell and the output
compared with that obtained from a reference cell.

Although the Golay cell in its standard form suffers from
microphony and a slow response time, it can be used over a broad
wavelength band from the visible to the microwave region, normally
determined by the cell window, and its performance approaches that of
an ideal room temperature detector.

3.5 Pyroelectric Detectors

3.5.1. Introduction

The pyroelectric effect has been known for several hundred years,
as a result of the studies of crystallographic structures. Although
many of the non-centrosymmetric classes of crystals exhibit a
spontaneous electric polarization effect, no external electric field
is normally observed. If the material is a conductor the mobile
charge carriers will assume a distribution which neutralizes the
internal dipole moment, but in a good pyroelectric material, which
will also be a good insulator, the extrinsic charge distribution is
relatively stable and even quite slow temperature changes in the
sample will produce changes in the lattice spacing and the dipole
moment, and since the stray surface charge will not be able to
respond rapidly to these changes an external electric field may now
be measured. Hence, the temperature coefficient of the dipole
moment can be determined and this is known as the pyroelectric
coefficient (Cady 1946).

It can be seen that one important difference between pyroelectric
devices and other thermal detectors is that the former will only
respond to the rate of change of temperature, rather than the actual
temperature rise as is the case for other types of thermal sensors.
Thus the pyroelectric device is used in an a-c mode at a frequency

sufficiently high to prevent stray charge from neutralizing the
effect before it is measured. This generally limits the low
frequency operation to approximately 1 Hz, and for a maximum output
signal the rate of change of the input radiation should be comparable
to the electrical time constant of the element. Although the
possibility of using the pyroelectric effect as a thermal detector
was proposed over forty years ago (Yeou Ta 1938), little progress was
made due to the lack of suitable materials. It is only in the last
twenty years that materials with sufficiently large pyroelectric
coefficients were discovered and a requirement for uncooled, rugged
thermal detectors with better performance than the thermistor
bolometers or rugged thermopiles emerged. It is now possible to
detect temperature changes of less than 10^{-6} C, which is comparable
to the Golay cell or a high sensitivity thermopile, but the detector
is much more robust.

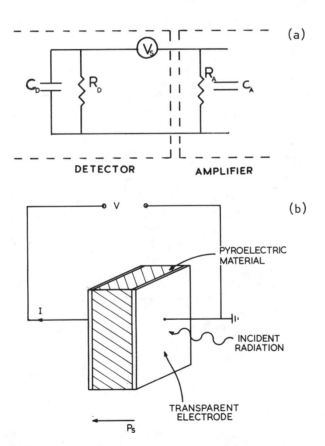

Fig. 3.4 Schematic representation (b) of a pyroelectric detector and
         its equivalent electrical circuit (a).

3.5.2 Detector Performance

The pyroelectric detector can be considered as a small capacitor
with two conducting electrodes mounted perpendicularly to the
direction of spontaneous polarization, as shown in Figure 3.4(b) with
its equivalent electrical circuit in Figure 3.4(a). Initially, the
pyroelectric material consists of a large number of separate domains
with differing directions of polarizations producing a net zero
effect over the whole slice. To orient the slice before use, the
material is heated and an electric field applied. Subsequently when
the detector is operated, the change in polarization will appear as a
charge on the capacitor and a current will be generated, the
magnitude of which depends on the temperature rise and the
pyroelectric coefficient of the material.

For a detector of area A, and pyroelectric coefficent p, if
radiation falls onto the device normal to the axis at a modulated
angular frequency of w, a temperature rise θ will be produced and a
voltage V developed across the detector, such that

$$V = wpAr\theta (1 + w^2\tau_E^2)^{-\frac{1}{2}} \tag{3.1}$$

where

$$\tau_E = rC \tag{3.2}$$

is the electrical time constant of the output circuit of the
detector.

The radiation absorbed by the detector causes its temperature to
rise thus changing its surface charge, hence to determine its
performance both the thermal and electrical characteristics must be
calculated. These are discussed in detail by Putley (1970).
Typically the surface charge is of the order of 10^{-16} coulombs
on a capacitance of the order of 10 pF. Consequently low noise high
impedance amplifiers are required for these systems, such as junction
field effect transistor (JFET) source followers.

In the high frequency limit the responsivity is given by

$$R = \frac{npA}{wHC} \tag{3.3}$$

where H is thermal mass of the detector.

As discussed in Section 2.8 the principal noise sources are
temperature, Johnson and amplifier noise.

3.5.3 Pyroelectric Materials

Many pyroelectric materials have been investigated for detector applications, however the choice is not an obvious one as it will depend on many factors including the size of the detector required, the operating temperature and the frequency of operation. As most pyroelectric materials are also ferroelectric, it is necessary to operate these devices at temperatures below their Curie temperature, as the polarization of pure materials when raised above this temperature reduces to zero and remains so upon cooling.

The principal materials used for pyroelectric detection are triglycine-sulphate (TGS) type materials, lithium tantalate (LT), strontium barium niobate (SBN), ceramic materials based on lead zirconate titanate, and more recently, the polymer films, polyvinyl fluoride PVF and polyvinylidene fluoride, PVF_2.

TGS is considered to produce the most sensitive detectors and has been used in the pyroelectric vidicon, unfortunately it is hygroscopic and relatively fragile, and its major disadvantage is its relatively low Curie temperature ($49^{o}C$), which limits its application. However Lock (1971) discovered that by doping the TGS with the amino acid L-alanine, an asymmetry in the hysteresis loop was produced and the material retained a preferred poling direction, thus eliminating the requirement to repole the material after heating to temperatures above the Curie point. The doped material also possesses a higher pyroelectric coefficient and a lower dielectric constant, and detectors with D* values of 2×10^{9} cm $Hz^{\frac{1}{2}}$ W^{-1} have been obtained at 10 Hz.

An alternative range of materials used for pyroelectric detectors are ceramics generally fabricated from lead zirconate titanate (PZT) or lead titanate. These materials can be produced in large areas, unlike the single crystal devices, generally by hot pressing techniques. They are robust and cheap and can be poled after processing, in any desired direction, by the application of a suitable electric field at an elevated temperature. Ceramic devices have D* values in the low 10^{8} cm $Hz^{\frac{1}{2}}$ W^{-1} range, the performance of these devices is comparable or better than lithium tantalate, except in the case of large area detectors.

The third category of materials used are the plastic films, (such as polyvinylidene fluoride). Their main advantage is that they can be produced in large areas requiring little processing, and are relatively cheap. However the performance is inferior to the other categories of materials, except for very large detectors operating at high frequencies.

To obtain a direct comparison between the various pyroelectric materials is very difficult as the detector area and operating

frequency will effect the performance, and account must also be taken
of the environmental operating conditions. Porter (1981) has
compared devices operating under different conditions, for detector
areas ranging from 100 mm^2 to 0.01 mm^2. In the case of the large
area devices, TGS and lithium tantalate appear to be the best devices
for all frequencies, except at very high frequencies (> 10 KHz) when
the polymer film devices begin to dominate. However for small
detectors strontium barium niobate (SBN) is preferred, for
intermediate area devices the performance of all devices is comparable
and other considerations become more important. These results only
show the trends as varying the detector parameters or the FET
amplifier could alter this situation. A summary of the properties of
various pyroelectric materials is given in Table 3.2.

With recent advances in the fabrication and production of cheap
devices their use for many applications has become commonplace.
Generally the performance of these detectors is attractive in the
frequency range 1-100 Hz, but falls off quite rapidly outside this
region.

3.5.4 Applications

Pyroelectric devices are now used in a variety of applications
which exploit two of the basic characteristics of the device (Putley
1981). Firstly, they only respond to changes in incident radiation
and are thus ideally suited for the detection of very small changes in
flux whilst operating with a large background level of incident
energy. The other important feature is their broad spectral band
response which has permitted their use for the detection of radiation
from the microwave region to X-rays.

One of the largest markets for these devices is for intruder
alarms, and several hundreds of thousands are produced each year.
The device operates on the principle that a constant radiation flux
is obtained from the scene under surveillance until an intruder
enters and the resulting change in flux is easily detected. The
device normally operates in the 8-14 μm band, which matches the
emission from objects close to ambient temperature, and the effects
of sunlight can be eliminated with a short wavelength filter. The
efficiency of the system can be improved by using a facetted mirror
to subdivide the scene and as the intruder moves through the beam a
low frequency signal will be produced. Generally these systems are
fabricated with either PZT ceramics or $LiTaO_3$ devices, both of
which can be produced in large quantities at relatively low cost.

A similar application is for fire detection, but generally the
devices operate at shorter wavelengths ~ 4 μm, and are triggered by a
fluctuating signal, corresponding to the typical flicker frequency of
flames in the range 5-40 Hz. As they must operate at relatively high
temperatures, $LiTaO_3$, which has a Curie temperature of 620° C, is

Table 3.2 Properties of Pyroelectric Materials

Material	Curie Temp °C	Pyroelectric Coefficient C cm^{-2} K^{-1}	Dielectric Constant	Thermal Conductivity W cm^{-1} K^{-1}	Specific Heat J g^{-1} K^{-1}	Density g cm^{-3}
TGS	49	3×10^{-8}	30	6.8×10^{-3}	0.96	1.69
LiTaO$_3$	618	6×10^{-9}	58		0.29	8.2
BaTiO$_3$	126	2×10^{-8}	160	9×10^{-3}	0.5	6.0
LiNbO$_3$	1190	4×10^{-9}	30			4.64
SBN	115	6×10^{-8}	380		0.4	5.2
PVDF	120	3×10^{-9}	10			
PZ	200	3.5×10^{-8}	250			

the preferred material, for this application. More recently this mode
of operation has been exploited for weather satellites (to determine
for example, the temperature profile of the atmosphere), and interplanetary probes, such as the pioneer mission to Venus. These
systems have used the more sensitive doped TGS detectors, for their
higher performance (Hamilton et al 1975).

Another important category of pyroelectric systems is the
spectrometer and one example of this is for the analysis of the
contents of milk. The Foss milk testing apparatus will measure the
fat, protein and lactose contents of milk samples.

Similarly the presence of a specific gas, for example CO_2, can
be determined by monitoring the absorption spectrum of an infrared
source transmitted through a gas sample. The source must be
modulated and the correct filter, corresponding to the absorption edge
included.

Pyroelectric detectors are also being used in conjunction with
infrared lasers, such as CO_2 (10.6 μm) and HCN (337 μm). For most
applications extreme sensitivity is not required, but for pulsed
systems the speed of response is vital, and special pyroelectric
devices have been developed with rise times faster than 1 nsec.
Stotlar et al (1980) have described a detector with a rise time of
about 30 psec, achieved by matching a strontium barium niobate, SBN,
device to a 50 ohm resistor. Roundy et al (1974) have produced
$LiTaO_3$ devices with similar response times.

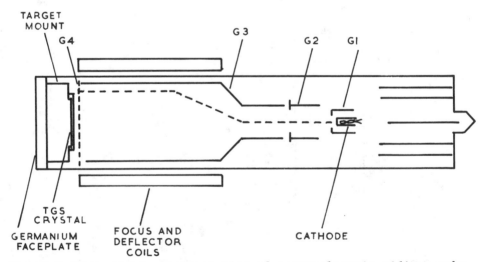

Fig. 3.5 Schematic representatation of a pyroelectric vidicon tube,
         showing the TGS pyroelectric crystal and the scanned
         electron beam readout.

The requirement for fast devices with high responsivity is not generally satisfied. As previously described the sensitivity peaks at low frequencies and will be several orders of magnitude down in the megahertz range.

One of the most important applications of thermal detectors, in particular pyroelectric devices is for infrared imaging. Originally an evapograph was employed in which the radiation was focused onto a blackened membrane coated with a thin film of oil. The differential rate of evaporation of the oil was proportional to the intensity of the radiation. The film was then illuminated with visible light to produce an interference pattern corresponding to the thermal picture. The performance was poor due to the very long time constant and the poor spatial resolution.

An improvement was found by scanning a single element detector such as a thermistor (Astheimer and Wormser 1959) or a pyroelectric element (Astheimer and Schwarz 1968), across the scene. Unfortunately as the response time of these devices is long real time imaging is not practical. A major advance was the development of the pyroelectric vidicon tube, which can be considered analogous to the visible television camera tube except that the photoconductive target is replaced by a pyroelectric detector and germanium faceplate. The concept was first proposed by Hadni (1963), and commercial pyroelectric vidicon tubes can now be readily purchased. A schematic of a tube is shown in Figure 3.5. It consists of a disc of pyroelectric material with a transparent electrode deposited onto the

Fig. 3.6 Thermal picture taken with a pyroelectric vidicon camera
       as shown in Fig. 3.7.

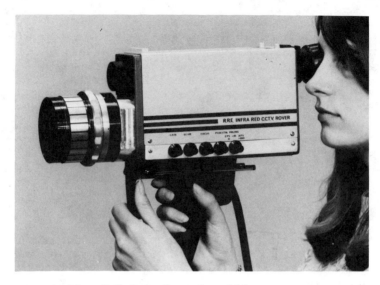

Fig. 3.7 Pyroelectric vidicon camera.

front surface. The charge output is then read by scanning an electron
beam across the reverse side of the target. The original tubes were
fabricated with TGS but better results have now been achieved with
deuterated TGS and TGFB material (Watton 1976, Stupp 1976). One of
the major problems with imagers using thermal detectors is due to the
thermal spread of the signal which reduces the spatial resolution,
however since the radiation input must be modulated to produce an
output from a constant thermal scene in any pyroelectric device this
problem is reduced. A further improvement can be obtained by
reticulating the target into 50 μm square areas, to limit the thermal
diffusion (Stokowski et al 1976, Yamaka et al 1976). Vidicon tubes
have now been produced with 0.2°C resolution in an image consisting of
100 TV lines and with reticulation the spatial resolution should
be increased further (Warner et al 1981). A typical thermal image
obtained with one of these devices is shown in Figure 3.6, and a
pyroelectric vidicon camera in Figure 3.7.

   Although good quality imagery has been obtained from the
pyroelectric vidicon, recent work has investigated the possibility of
using a solid state read out and interfacing two dimensional arrays of
detectors directly with a silicon circuit. This should improve the
temperature resolution of the system and produce a more robust and
lighter imager. Each discrete detector is integrated with a silicon
CCD, thus allowing the possible access of individual cells for
quantitative measurements (Watton et al 1982, 1983). Several
pyroelectric materials have been studied but the most commonly now
adopted are $LiTaO_3$, PVF_2 and PZT ceramics. Small arrays have been
fabricated and imaging demonstrated. The possibility of producing

very large arrays, 10^4 - 10^5 elements, enabling the removal of all scanning components, is a major attraction of this system.

The pyroelectric detector is gradually replacing other thermal detectors for applications which require room temperature operation, since they can offer an attractive combination of performance, reliability and cheapness. The use of ceramic devices for burglar alarms is a large volume production area for the cheap devices, whilst the modified TGS detectors are available for the higher performance systems, such as satellites. The pyroelectric vidicon has produced good quality thermal imaging with an uncooled detector, and is comparatively cheap, and it is hoped to eventually produce a pyroelectric CCD camera for imaging applications.

The possibility of using these devices for submillimeter range operation has recently been proposed by Hadni et al (1978), who suggested running the devices at cryogenic temperatures. However, this must be compared to the alternatives of liquid helium cooled bolometers or extrinsic photon devices.

CHAPTER 4

PHOTOEMISSIVE DETECTORS

4.1 Introduction

The principle of photoemission was first demonstrated in 1887 when Hertz discovered that negatively charged particles were emitted from a conductor if it was irradiated with ultraviolet. Further studies revealed that if an alkali metal electrode was used, this effect could be produced with visible radiation, (Elster and Geitel 1889). Although these effects could be demonstrated reproducibly, no satisfactory explanation was offered, until in 1905, Einstein proposed his theory of photoemission, and these effects were then explained in terms of electron emission induced by the incident radiation.

Following this initial period many different materials were studied and their spectral response and threshold wavelengths investigated. However, the use of the effect for practical systems was limited by the very low quantum efficiencies available ($< 10^{-4}$ electrons/incident photon), but, in 1929 the situation was significantly altered with the discovery of the silver-oxygen-caesium (Ag-O-Cs) photocathode by Koller (1929, 1930) and Campbell (1931). This material had a quantum efficiency two orders of magnitude above anything previously studied, and consequently a new era in photoemissive devices was inaugurated. However as the fundamental mechanism of this process was still not sufficiently understood to enable improved materials to be modelled, the studies were mainly empirical, aimed at producing materials which extended the long wavelength sensitivity into the near infrared.

It was not until it was realized that these materials were semiconductors and, in parallel, progress had been made in solid state physics that photocathodes and their behaviour were fully understood.

45

Emphasis then increased on practical applications of the phenomenon, and devices such as the photomultiplier and iconoscope were produced.

4.2 Principle of Operation

The process of photoemission from any material can be considered in three stages, namely: (1) the excitation of the photoelectron, (2) its diffusion to the emitting surface, (3) the escape of the photoelectron into the surrounding vacuum.

In any material the electrons are bound by the ionization energy to the lattice, and any electron at the surface which is excited to an energy level greater than the ionization energy will have a high probability of escape. In order to excite an electron in the lattice an incident photon must be absorbed. In the case of a metal the photon reflectivity is high in the visible and near infrared region of the spectrum, and the energy loss of any excited carrier will be rapid. Thus, the quantum efficiency of any metal photocathode will be low.

An improvement over a metal photocathode is obtained by using a semiconductor which will have a higher photon absorption efficiency and longer relaxation times for energy losses, thus increasing their quantum efficiency; and their lower ionization potential allows longer wavelength operation.

Once created, the photoelectron within the material must be transferred to the surface. The probability of this hot electron reaching the vacuum interface depends on the mean free path and the energy loss processes for the electron.

In a metal, electron scattering is the dominant process and due to the high density of free carriers the mean free path will be short. Consequently, only those electrons created within a few atomic layers of the surface can escape. However, for the semiconductor this loss mechanism is negligible and the two methods by which a photoelectron will lose energy are by electron-hole pair production (impact ionization) and by lattice scattering (phonon production).

The energy loss per scattering event is quite small, typically 0.01 eV, and the mean free path is 30-40 A^o, hence in the absence of pair production electrons produced at depths of several hundred angstroms can have sufficient energy for emission. However if an electron has an energy above a certain threshold limit, E_{Th}, it can lose energy by creating an electron-hole pair. This threshold energy is generally several times larger than the energy gap, E_G, of the semiconductor, and the mean free path will be of the order of 10 - 30A^o.

Finally, the photoelectron must have sufficient energy when it reaches the interface to be able to overcome the surface potential.

For a metal the work function, ϕ, is defined as the energy difference between the Fermi level and the minimum free energy of an electron at rest within the vacuum, Evac = 0. For all metals this is at least 2 eV, consequently the long wavelength response has a threshold at approximately 6000A$^{\circ}$. In the case of a semiconductor, doping the material will alter the position of the Fermi level and a better parameter than the work function is the electron affinity, E_A, which is defined as the difference in energy between the bottom of the conduction band and the vacuum level, which will be a constant of the material, independent of the impurity content.

Three different conditions can be considered, depending on the relative sizes of E_{Th} and E_A, as shown in Figure 4.1.

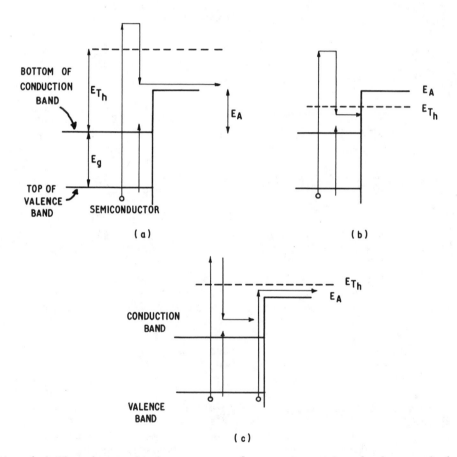

Fig. 4.1 The photoemission process for a conventional photocathode. Three different conditions are considered depending on the relative value of the threshold energy, E_{Th}, and the electron affinity, E_A.

(a) If E_{Th} is large compared to E_A, an electron will have a high probability of escape, even after producing a secondary electron-hole pair.

(b) For materials with E_A larger than E_{Th}, any electron which has sufficient energy to escape will also be able to create an electron-hole pair, and due to the very short mean free path of pair production, the only electrons with a chance to escape will be those generated very near to the surface.

(c) If E_{Th} is only slightly larger than E_A, electrons will only be able to escape if they have an energy greater than E_A but less than E_{Th}, hence the quantum yield will be moderately low.

Consequently the minimum energy required for a photon to produce photoemission will be $(E_G + E_A)$, unless the density of electrons in the conduction band was sufficiently high to allow photon absorption, but this is seldom the case. If $(E_G + E_A)$ is less than 3 eV, the emission will be in the visible part of the spectrum, however by careful choice of the semiconductor material the threshold has now been extended into the near infrared.

Thus it can be seen that high quantum efficiency and long threshold wavelengths will only be achieved with semiconducting photocathodes. Examples of the most important photocathodes and their properties will be discussed in the following sections.

4.3 Conventional Photocathodes

4.3.1. Metal Photocathodes

In 1888 Hallwachs discovered that if a negatively charged zinc electrode was illuminated with ultraviolet radiation it would lose its charge, subsequently it was shown that alkali metal cathodes produce photoemission with visible light. Although much work was carried out on measuring the properties of these photocathode materials, some of the results obtained before the development of modern vacuum technology are questionable, as small quantities of absorbed gases on the surface can cause very large variations in the work function. A summary of the more important results obtained is given by Sommer (1968) in his review of photoemissive materials.

The efficiency of the photoemissive material is also controlled by the optical reflectivity and the absorbtion coefficient, and as for most metals the reflectivity is high, typically 90-99%, their quantum yield will be low, thus they have now been replaced by semiconductors, in which the photon absorption is more efficient and the relaxation time for energy losses longer.

4.3.2. Semiconducting Photocathodes

To simplify the identification of photocathodes, each combination has been given an internationally agreed S number, as specified by the Electronic Industries Association (1964), and a list of the more important photocathodes with their S numbers and characteristics are shown in Table 4.1, and typical spectral sensitivity curves in Figure 4.2.

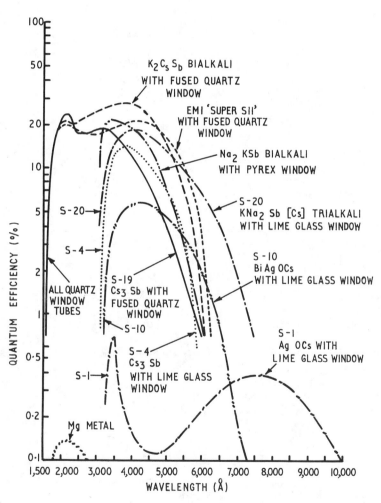

Fig. 4.2 Typical spectral sensitivity curves of some commonly used photocathodes.

Table 4.1

Registered S Numbers for Specified Photosensitive Surfaces and
Window Materials

Spectral Response Number	Photoemissive Material	Window Material	Wavelength off peak Response ($\overset{\circ}{A}$)
S-1	Ag-O-Cs	Lime glass	8,000
S-4	Cs_3Sb	Lime glass	4,000
S-5	Cs_3Sb	UV transmitting glass	3,400
S-6	Na	(Unspecified)	
S-9	Cs_3Sb (Semitransparent)	Lime glass	4,800
S-10	Bi-Ag-O-Cs (Semitransparent)	Lime glass	4,500
S-11	Cs_3Sb (Semitransparent)	Lime glass	4,400
S-12	CdS	Lime glass	
S-13	Cs_3Sb	Fused quartz	4,400
S-19	Cs_3Sb	Fused quartz	4,440
S-20	Sb-K-Na(Cs) (Semitransparent)	Lime glass	4,200
S-21	Cs_3Sb	UV transmitting glass	
S-24	Na_2KSb	Lime glass	

The S-1 Ag-O-Cs Photocathode

The first semiconductor photocathode was developed by Koller in 1929 (Koller 1930), and consisted of a layer of caesium on oxidized silver, it is sensitive throughout the visible and near infrared regions of the spectrum. As well as being one of the first photocathode materials, the S-1 is still the only conventional semiconductor with useful response in the near infrared, out to approximately 1.2 μm, and the ultraviolet region, down to 0.3 μm. However, the quantum efficiency is low throughout the region, typically 1/2%, and the cathodes show long term decay after storage and rapid deterioration under conditions of high illumination. Typical responses of 30 μA/lm can be achieved in the ultraviolet region, and dark currents of 10^{-13} to 10^{-11} amps cm^{-2} are obtained. These cathodes are generally fabricated by depositing a thin layer of silver on a glass carrier which is then oxidized by a glow discharge, or r.f. heating. The layer is then sensitized with caesium vapour and heated to about $130^{\circ}C$.

Unfortunately this fabrication procedure is quite critical and the yield low, thus increasing the cost of S-1 devices, hence they are not widely used if an alternative is available. Also, the exact mechanism for photoemission is still not completely understood, although a possible band structure has been proposed by Zwicker (1977). It is probable that the emission from this photocathode is a two-step process with absorption taking place in the silver, which is most sensitive in the infrared region and emission occurring through the low electron affinity Cs_2O material.

Antimony - Caesium (CsSb)

The caesium - antimony photocathode is one of the most widely used due to its high quantum efficiency. It was originally discovered by Gorlich in 1933, (Gorlich 1936), and has since been studied extensively. The fabrication is relatively simple: an antimony film is evaporated onto a faceplate, which is then activated by evaporating caesium onto the antimony which is maintained at an elevated temperature (Sommer 1968). After careful cooling to room temperature, a highly stable film of Cs_3Sb is formed with an energy gap of 1.6 eV and an electron affinity of 0.45 eV. The threshold voltage is approximately 2 eV.

The band structure of this material has been investigated thoroughly and the shape of the surface band bending predicted by Wooten and Spicer in 1964. This is shown in Figure 4.3.

The sensitivity of these tubes can be increased if a small amount of oxygen is introduced and the cathode reheated. Depending on the window material and if the cathode is deposited onto an opaque or semi-transparent substrate the Cs_3Sb tubes have been given a range of

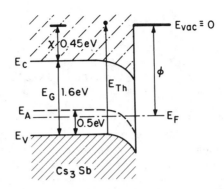

Fig. 4.3 The approximate band structure of a CsSb photoemissive layer
(Sommer 1968).

S numbers including S-4, 5, 11, 13, 17 and 19. Consequently, due to
its relatively low cost and high stability it is now used frequently.

Other Materials

The bismuth-silver-oxygen-caesium, S10, cathode has a quantum
efficiency similar to the Cs_3Sb, but its spectral response closely
follows that of the human eye, which is important for many
applications. A large number of multialkali antimonide photocathodes
have been produced. Na_2KSb has a peak sensitivity in the blue region,
and although its quantum yield is somewhat lower than other
multialkali antimonides it can be used at temperatures of up to 150^oC.
However, by introducing a small amount of caesium the response can be
extended into the near infrared, and when used with a lime glass
window this is the S-20 photocathode. Sensitivities of up to 500μA/lm
have been reported for this device with dark currents of 10^{-15}
amps cm^{-2}. For shorter wavelength applications the oxidized
$K_2 CsSb$ (0) cathode, which peaks in the blue region is used,
especially to detect nuclear radiation in a scintillation counter.

These multialkali cathodes generally have high quantum
efficiencies but are difficult to prepare, consequently their use has
generally been limited to more specialized photomultiplier and image
intensifier tubes. For general purpose use the Cs_3Sb cathode is
employed, or the Ag-0-Cs if response in the near infrared is required.
A review of the most popular photocathodes, describing their
fabrication methods and sensitivities has recently been published by
Ghosh (1982).

4.4 Negative Electron Affinity Devices

To improve the photoemission yield and extend the wavelength
sensitivity it can be seen from the discussion on classical

photoemitters that the electron affinity should be reduced, or ideally become effectively negative.

It had been known for many years that by depositing a layer of caesium onto many materials the electron affinity could be reduced, but not made negative. However, by considering the band structure of semiconductor materials, Spicer (1958, 1960) was able to predict that by using a heavily doped material an effective negative electron affinity cathode could be fabricated. Eventually, in 1965, Scheer and Van Laar, achieved this with gallium arsenide which was heavily doped p type (of the order of 10^{19} cm^{-3}) and coated with a layer of caesium. This produced a photocathode with a high quantum yield up to the cut-off wavelength of the material. From these results it was concluded that the emission was independent of the surface properties, and hence that the vacuum level of the semiconductor must have been lowered below that of a free electron in the conduction band of the crystal, such that any electron excited in the bulk, if it reaches the surface of the material without recombining, will be freely emitted. Consequently this is a cold electron emission device, compared to the hot electron process required for a classical photoemitter. Any photocathode which exhibits this property is known as a Negative Electron Affinity, NEA, photoemitter.

The band structure for such a device is shown in Figure 4.4, for p type material. It can be seen that although an electron at the surface will not have an energy greater than E_A, there is a region in the bulk for the p type material where the energy of an electron at the bottom of the conduction band will exceed the vacuum potential, a situation which cannot occur with a conventional photoemitter. Thus these devices should be referred to as exhibiting effective negative electron affinity.

However, by increasing the doping level the region in which the band bending occurs can be made small with respect to the electron escape depth and the optical absorption length, which is the normal situation for NEA devices, hence they can be considered as negative electron affinity emitters and are fabricated from p type degenerately doped crystals.

The mechanism and band structure of various NEA photoemitters has been reviewed by Bell (1973) and Zwicker (1977).

The first devices were produced on GaAs (Scheer and Van Laar 1965) and GaP (Williams and Simon 1967) substrates, with a monolayer of caesium and sensitivities of approximately 500µA/lm were obtained. A significant improvement was achieved by replacing the simple layer with a thicker layer of caesium and caesium oxide, this further reduces the electron affinity as shown in Figure 4.5 and increases the sensitivity to 2000µA/lm (James et al 1971[1], Olsen et al 1977). The threshold energy for these photocathodes is determined by the band gap

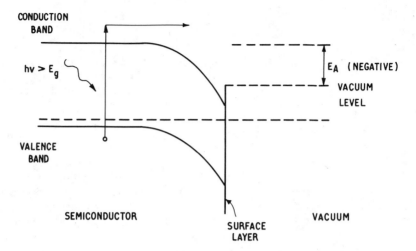

Fig. 4.4 The photoemission process and energy band structure for a p
         type negative electron affinity photocathode.

energy, thus to extend the coverage into the near infrared, direct
band gaps of less than 1.2 eV are required which is approximately
equivalent to 1 μm cut off wavelength. However, the overall quantum
efficiency at shorter wavelengths will be reduced, as shown in Figure

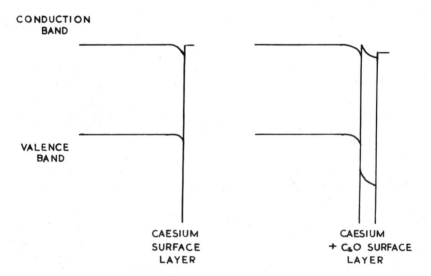

Fig. 4.5 The energy band structure of a GaAs photocathode with a
         caesium surface layer and a caesium and caesium oxide
         layer, which increases the sensitivity.

4.6.   The (III-V) ternary and quaternary alloys have the advantage
that their band gaps can be varied by adjusting their composition,
consequently the wavelength response of photocathodes fabricated from
these materials can be adjusted for a particular application.
However, the long wavelength response cannot be extended beyond a
point at which the bottom of the conduction band lies below the top of
the interfacial barrier, as shown in Figure 4.7.

Also, although the surface electron affinity can be reduced the
NEA efficiency drops as the energy gap is reduced thus the height of
the interfacial discontinuity limits the wavelength response. The
most important alloys are the quaternary InGaAsP (Escher et al 1976)
and its related ternaries InAsP (James et al 1971[2], Antypas et al
1972) and InGaAs (Fisher et al 1972). Although by reducing the energy
gap, response beyond 1.1 µm can be obtained the quantum efficiency
drops rapidly, due to the interface potential barrier between the bulk

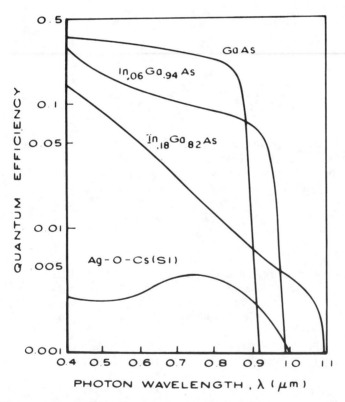

Fig. 4.6 The quantum efficiency of three members of the InGaAs family
         of photocathodes, including GaAs, showing the drop in yield
         as the band gap is lowered to extend the wavelength response.
         The S-1, Ag-Cs-O, curve is included for reference (Martinelli
         and Fisher 1974).

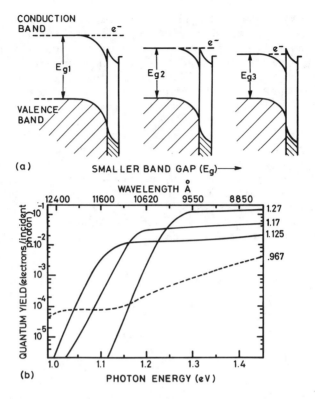

Fig. 4.7 (a) The schematic band diagram for a variable energy gap III
- V material. (b) The quantum yield of $InAs_xP_{1-x}$ alloys for
various compositions, the energy gap is given on the right-
hand abscissa for each composition. The dashed curve
illustrates the drop and change in shape of the yield curve
when the conduction band minimum drops below the interfacial
barrier (Spicer 1977).

material and the caesium oxide layer. Quantum efficiencies of 9% and
5.4% for InGaAsP and InAsP cathodes have been reported. These are
approximately two orders of magnitude greater than obtained from
classical photoemitters, this is illustrated in Figure 4.8.

     One important application for these devices is in conjunction with
a Nd-YAG laser system, which operates at 1.06 µm.

     When investigating NEA activated materials for infrared operation,
silicon was thought to offer many advantages due to its narrow band
gap of 1.1 eV (Martinelli 1970). It produces a higher yield near 1 µm
than InGaAs, due to its long electron lifetime, and hence long carrier
escape depth (10µm), however it has a small absorption constant owing
to its indirect band gap. Unfortunately it also has a very high

thermionic dark current, 10^{-9} amps cm^{-2} at room temperature, which is
many orders of magnitude above that of other NEA emitters.   This
problem combined with the more critical activation required for a
silicon NEA cathode has limited its use.

One method of extending the wavelength of operation of a
photocathode is that of field assisted emission, in which an external
field is applied to assist the emission of photoelectrons from the
surface of the material. This effect was used with conventional
photocathodes before the production of the first NEA devices.   Simon
and Spicer (1960) demonstrated that the emission from a germanium
surface could be extended from approximately 0.8μm to 2μm. However,
unfortunately only very low efficiencies have been obtained.

4.5   Photoemissive Devices and Applications

There are several applications for which photemissive devices are

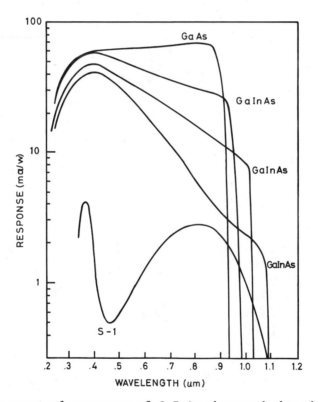

Fig. 4.8 The spectral response of GaInAs photocathodes showing that
the response falls as the wavelength is increased, due to the
interface potential barrier between the bulk material and the
caesium oxide layer (Thomas 1973).

ideally suited, these include the detection of very low intensity
signals, high speed pulses of radiation and for uses requiring very
high spatial resolution, such as imaging.

The most important features of photoemissive devices compared to
other photodetectors are that fast, high gain, and low noise internal
amplification can be incorporated within the detector by use of an
electron multiplier, and secondly uniform, large area detectors can be
fabricated routinely, thus allowing direct view imaging devices to be
constructed. For high spatial resolution applications image tubes can
be integrated with a scanned electron beam read-out.

The major problem with these devices is their limited wavelength
response in the infrared region. Devices were not available which
responded beyond approximately 1μm until the recent advances with NEA
photocathodes which allow detection out to 1.5μm. At the short
wavelength end, the limiting factor has been the suitability of a
window material. Lithium fluoride allows detection down to 0.1μm,
below that the devices have to be operated without a window, but in
this mode they have been used to detect X-rays.

4.5.1 Photodiode

The simplest form of photoemissive detector is the vacuum diode,
which consists of a photocathode, sensitive to the radiation
wavelength of interest, and a positively biased anode which collects
electrons emitted from the photocathode, contained within an evacuated

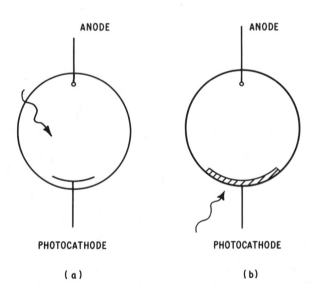

Fig. 4.9 Photodiode tube with (a) an opaque photocathode and (b)
semitransparent photocathode.

encapsulation with a suitably transparent window. Generally the anode
consists of a thin metal rod, placed at the centre of the tube, as
shown in Figure 4.9(a). Alternatively a semitransparent cathode can
be deposited on the inside surface of the glass envelope, as shown in
Figure 4.9(b), which will prevent any problems that might arise due to
obscuration of the cathode by the anode. The thickness of the layer
must be accurately controlled to optimize the photon absorption and
the photoelectron emission, this normally corresponds to a thickness
typically of less than 100 A$^\circ$.

These devices are usually operated such that all the emitted
electrons are collected by the anode, in this mode the current varies
linearly with the intensity of the incident radiation and is almost
independent of the applied voltage. This generally corresponds to an
anode voltage of between 50 and 150 volts positive with respect to the
cathode, and the current is normally limited to a few microamps, see
Figure 4.10. If the voltage is set too high, it is possible that any
residual gas in the tube could be ionized, thus reducing the response
linearity.

Another important characteristic of these devices is their very
high frequency response, in the region of 100MHz, due to the small
spread in transit times between the photocathode and anode, of
electrons emitted with different velocities.

The short term stability of these tubes is excellent, consequently
they have been employed for exact measurements of light levels, but
their lifetime is reduced if they are operated under high illumination
conditions, and the sensitivity falls. However, this can be restored
partially or completely if the tube is stored in the dark. The
overall life of the tube is related to the reciprocal of the current
passing through it.

The most important noise sources in the phototube are shot noise
and Johnson noise. The shot noise is due to fluctuations in the rate
of arrival of electrons at the anode, and the root mean square voltage
is given by

$$(\overline{\delta v^2})^{\frac{1}{2}} = R(2ei\Delta f)^{\frac{1}{2}} \tag{4.1}$$

The Johnson noise is developed across the anode load resistor such
that

$$(\overline{\delta v^2})^{\frac{1}{2}} = (4kTR\Delta f)^{\frac{1}{2}} \tag{4.2}$$

Thus by adjusting the value of the load resistor the major noise
source can be determined, if a low resistor value (100 Kohm) is
chosen the Johnson noise will dominate and a good frequency response
will be achieved. Alternatively for low level signals and if a lower
response time can be tolerated, much larger value anode resistors

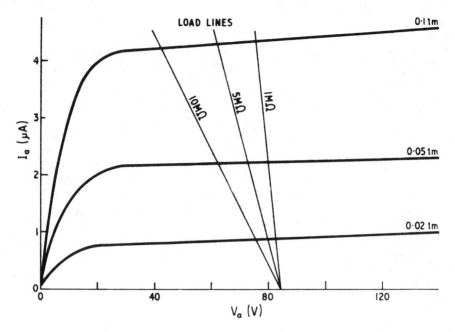

Fig. 4.10 The anode current/anode voltage characteristics and load
         lines for a Mullard vacuum photodiode.

(100 megohm) are selected.

As previously discussed there is no internal amplification within
the vacuum tube, but by the introduction of a gas at low pressure,
amplification factors of typically ten can be obtained. To prevent
contamination of the cathode material only inert gases are used, and
most frequently argon is selected. The principle of operation is
similar to the vacuum tube, however, the photoelectrons are now
capable of producing additional electrons by impact ionization of the
gas molecules. The amplification increases as a function of the anode
potential, but is limited by the onset of spontaneous gas discharge.

The frequency response of a gas filled tube will be limited to a
few thousand hertz due to the increased transit times caused by the
collision processes within the gas, and also the movement of the
heavier positive ions in the tube.

Photodiode tubes are capable of being fabricated with all types of
photocathodes, however their use is generally limited to the detection
and measurement of relatively large signals, consequently Cs_3 Sb and
Ag-O-Cs cathodes are used almost exclusively for visible radiation.
In the ultraviolet region the alkali halides and tellurides are the
most suitable.

Due to the non linearity of these devices with the incident radiation intensity, and the critical dependence of the gain on the anode voltage, gas-filled phototubes are unsuitable for photometry applications. However, they have been used frequently in film projectors to read the sound track on films. They have also been used recently to monitor the flame when natural gas is burnt, to ensure that no .unburnt gas escapes.

4.5.2 The Photomultiplier

Small values of internal amplification can be obtained by using a gas filled phototube, but, to achieve values of greater than a million, an electron multiplier is incorporated in the vacuum tube, such a device is known as a photomultiplier. The photoelectrons emitted by the cathode are accelerated by an electric field towards the first plate of the electron multiplier, known as the first dynode. The electron multiplier consists of a series of dynode plates which are coated so that they have a high secondary emission coefficient; consequently several secondary electrons are emitted from the surface of the dynode for each impinging electron. The plates are biased such that emitted electrons are accelerated to each succeeding dynode in the chain, where further amplification occurs. The electrons are finally captured by the anode. A photomultiplier tube generally contains between ten and fifteen dynode plates and an electric field of approximately 100V is maintained between successive plates. With this arrangement electron gains of up to 10^9 have been measured.

The characteristics of a photomultiplier are similar to those of the vacuum tube, but the advantages include the improved noise performance and faster response times compared to vacuum tube systems with external amplification. The noise improvement is due to the lower dark current and residual noise at each dynode than the photocathode, thus limiting the overall noise to either photon noise or residual cathode dark current noise, in contrast to most systems operating with external gain where the major noise contributions arise from the first stage of amplification.

The response time of a photomultiplier is determined by the spread in times of flight of the electrons between each stage, and with careful design of the dynode chain, bandwidths of 100MHz can be obtained.

The photocathode generally used in these devices is a semitransparent photoemissive material deposited onto the inside surface of the end face of the tube. This allows direct coupling of the source and detector, thus reducing any light loss. The choice of material for the photocathode is determined by the wavelength of operation. A high quantum efficiency is required in this region, with a low thermionic emission to reduce the dark current contributions, and finally the cost of the photocathode is often an important consideration. The short wavelength limit is usually determined by the phototube window material; lithium fluoride operates down to 1050 A°. But to achieve response at shorter wavelengths it is necessary to operate the device without a window, which is possible for some applications, such as in space. The recent developments in photocathodes employed in photomultiplier tubes and a summary of their sensitivities is given by Yokozawa (1982).

The choice of material to fabricate the dynodes in the electron multiplier, is determined by its secondary emission yield and stability and must normally exhibit lower dark current and residual noise than the cathode. Typically, secondary emission coefficients δ, of between three and six are obtained. Thus for a photomultiplier with n dynodes, a transfer efficiency between each dynode of g, and a collection efficiency between the photocathode and first dynode of f, the overall amplification of the tube, G, is given by

$$G = f(\delta g)^n \qquad (4.3)$$

There is normally an upper current limit for the dynode, and with limits on anode heating the stable gain is generally limited to 10^7, although for pulsed operation gains of 10^8 can be achieved.

The secondary emission process is dependent on the incident energy of the primary electron, as the number of electrons generated with sufficient energy to escape from the surface increases. However, with energies of greater than a few hundred volts the secondary electrons will tend to be produced at too great a depth in the material to escape. Typical emission curves for several materials as a function of incident electron energy are shown in Figure 4.11. Generally good photoemitters are good secondary electron emitters, consequently several of the classical materials are often chosen. The figure also shows the emission from an NEA material, GaP(Cs), which has a gain much larger than any of the conventional materials (Simon et al 1968), and increases linearly with applied voltage up to several thousand volts. The advantage of using materials with these higher secondary emission coefficients is that tubes with gains of 10^7 can be produced with only five stages, and these are now commercially available.

The three most popular materials selected for photomultiplier dynodes are caesium antimonide which offers a high gain at relatively low applied voltages, but has a poor high temperature stability; oxidized silver-magnesium alloy, which can be operated at higher currents and temperatures although requiring increased applied voltages; and finally oxidized copper containing about 2% beryllium, which is very similar in performance to the AgMgO system, but is somewhat easier to fabricate.

Generally the alloy films can be deposited onto the dynode plates

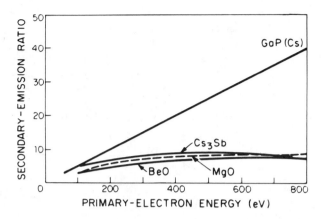

Fig. 4.11 The secondary emission ratios of several classical and one
         NEA electron multiplier dynode materials for a normal range
         of acceleration voltages (Krall et al 1970).

before the tubes are assembled, as they are stable in air, however the
caesium activation must be performed after evacuation.

As an alternative to coated dynodes, thin plates have been
constructed which allow the secondary electrons to be emitted from the
back of each plate. Potassium chloride has been used, and gives a
stage gain of about four, with an inter dynode voltage of
approximately 5kV, unfortunately the gain decreases with use (Blatner
et al 1965). More stable operation has been achieved using magnesium
oxide, but the gain is slightly lower.

The construction of the tube and dynode configuration should
ensure that all the electrons emitted by one dynode strike the
succeeding one and thus do not miss any stage of amplification, nor
should they collide with any other part of the tube. Many different
designs have been used to achieve this and some of the more popular
electrostatic constructions are shown in Figure 4.12. The tubes can
be considered as focussed or unfocussed systems. In the venetian
blind system each dynode consists of a number of small plates mounted
at 45° to the axis of the tube, the plates of each succeeding dynode
slope in opposite directions. To capture any low energy secondary
electrons a fine grid is mounted in front of each dynode. This system
exhibits very stable gain but poor response times. A similar
arrangement is used in the box and grid system, which is often used in
tubes of smaller diameter and has the highest dynode efficiency, but
tends to saturate at high anode currents, due to space charge effects.
In both of these tubes there is no focussing of the secondary
electrons between stages, but if the plates are curved a focussing
field will be obtained and the electrons will be directed towards the

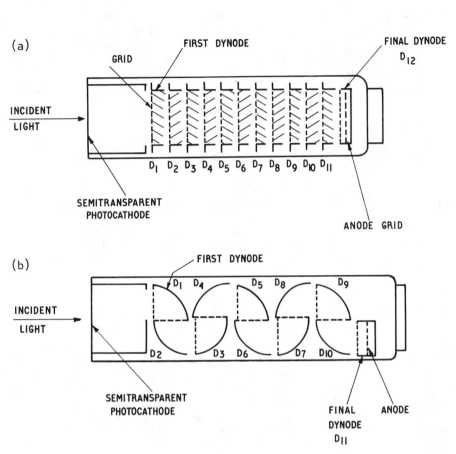

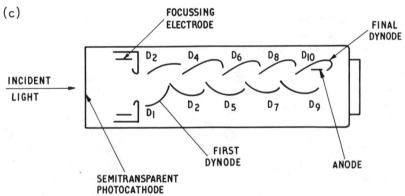

Fig. 4.12 Photomultiplier tube constructions (a) venetian blind system
          (b) box and grid system (c) focussed dynode linear chain.

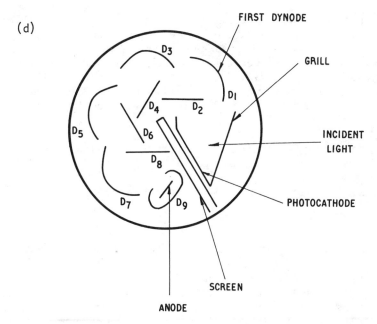

Fig. 4.12 (d) circular cage focussed dynode system.

centre of each dynode. Generally the electric field strength at the
dynode surface is greater in this type of tube than the unfocussed,
consequently the spread in transit times will be less, and output
pulses with rise times of less than 1 nsec have been achieved.

Focussed tubes can be constructed in the conventional linear
manner or more simply in a circular fashion, which enables a very
compact tube using an opaque photocathode to be assembled.

The electrons are finally collected at the anode, which often
consists of just a thin wire or grid mounted close to or inside the
final dynode, thus enabling any strong capacitance effects to be
minimized.

Inherent in any photomultiplier tube is the dark current, which is
the current collected at the anode when no illumination is incident on
the photocathode. Consequently it determines the lowest light levels
which can be detected. The dark current is due mainly to thermionic
or electric field emission from the photoemissive surfaces, in
particular the photocathode and first dynode, which is amplified by
the tube in the normal manner. The dark current contributions can be
reduced in several ways, for example, if the incident radiation is
focussed on to a small area photocathode, the signal to noise ratio
will be improved with respect to that obtained from a tube with a

large area photocathode. Alternatively, if the photomultiplier is cooled to approximately $-20^\circ C$ for conventional tubes or $-80^\circ C$ for an NEA device it is possible to reduce the thermionic dark current by an order of magnitude. A second source of the dark current is due to high energy particles, such as muons and beta particles in the background radiation causing scintillation in the tube window, and from stray electrons producing electroluminescence in the glass envelope.

The noise mechanisms effecting both signal and dark currents in a photomultiplier tube are similar to those discussed for the photodiode tube in the previous section, however, with the internal amplification of the tube, G, the shot noise will now increase to

$$(\overline{\delta v^2})^{\frac{1}{2}} = GR(2ei\Delta f)^{\frac{1}{2}} \tag{4.4}$$

The Johnson noise in the anode resistor will be unaltered.

Alternative designs of tubes have been considered using magnetically focussed multipliers (Heroux and Hinteregger 1960) and channel multipliers (Goodrich and Riley 1962). These employ a resistive film in place of the conventional dynode chain.

The concept of using a continuous dynode electron multiplier was first suggested in 1930 by Farnsworth, although their adoption took another 30 years. The multiplication in this type of tube is obtained from a continuous film deposited onto an insulator and biased such that an electron striking one end of the film generates secondaries which are accelerated along the tube and restrike the film further along its length. Hence the biasing arrangements are much simpler than in the conventional dynode system.

The Bendix magnetic multiplier uses a thin dynode film of tin oxide and antimony mounted opposite a similar parallel strip known as the field strip. The cathode is mounted at one end of the dynode and a grid placed above it, connected to the field strip, as shown in Figure 4.13.

The two films are biased typically to 2kV, but in such a manner that the potential at any point on the field strip is approximately 250V positive with respect to the corresponding point on the dynode film. A magnetic field is applied which causes the secondary electrons to spiral as they move along the tube and the electric field between the two strips causes them to collide with the dynode chain many times during their transit along the dynode, thus producing gains of up to 10^9.

This type of multiplier can be used with a variety of photo-cathodes depending on the application, and is generally small and rugged thus it is ideally suited for space systems.

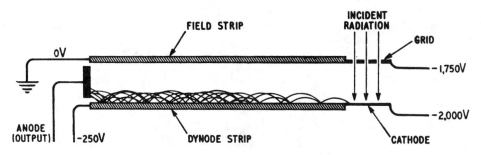

Fig. 4.13 Bendix magnetic electron multiplier.

An alternative method of obtaining high electron gain is by use of
the channel electron multiplier which consists of a hollow glass tube,
typically 1mm in diameter, coated with a resistive film which has a
high secondary emission coefficient. An electric field is applied
across the film and the electrons are multiplied as they accelerate
along the tube, and collide with the film, as shown in Figure 4.14(a).

The electron gain of the channel depends on the applied voltage,
the ratio of the channel length to the diameter and the secondary
emission characteristics of the channel surface. The gain of the tube
is up to 10^8 for 3kV applied field.

As the gain is determined by the ratio of channel length to
diameter and not the overall size, the dimensions can be scaled down
without affecting the performance. Thus, large bundles of these
multipliers can be grouped together to form a microchannel plate (MCP)
as shown in Figure 4.14(b). Each channel generally has a diameter of
between 10µm and 25µm and a length of 0.5mm, it is then possible to
incorporate 2.5 million of them in a 25mm plate. They are operated at
approximately 1kV with a typical amplification of 3000.

However, the gain of these tubes is limited by ionic feedback,
which arises due to electron impact ionization of the residual gas
molecules. These positive ions are then accelerated back down the
tube where they may gain sufficient energy to generate secondary
electrons if they collide with the wall.

To overcome this the walls of the tubes can be coated with a thin
aluminium film, typically 30A°thick , which acts as a barrier to the
positive ions, resulting in a long stable life for the photocathode,
although there is some evidence that the photoelectron collection
efficiency of the MCP can be reduced to the limitation imposed by the
open area ratio (Oba and Rehak, 1981). Alternatively, single channel
multipliers can be constructed in a circular configuration as
illustrated in Figure 4.15. Any ions now formed will only be able to
move a short distance down the tube before striking the wall, and

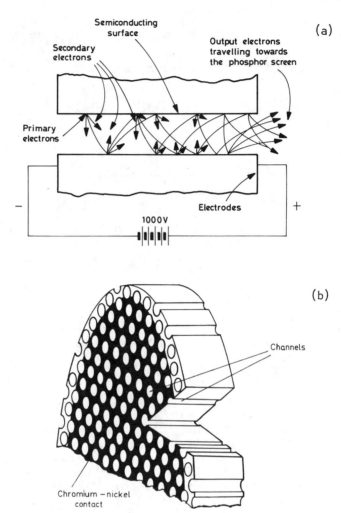

Fig. 4.14 (a) Operation of a single channel electron multiplier and
         (b) schematic representation of an electron microchannel
         plate.

hence will not have sufficient energy to generate electrons.   The
operation and construction of these tubes has been discussed more
comprehensively by Adams and Manely (1966) and Dhawan (1981).

    Similar gains can be attained with these devices as for the
magnetic multipliers previously discussed, but sacrifices in speed are
made for the sake of structural simplicity, although the tubes are
still a factor of two better than conventional photomultipliers.

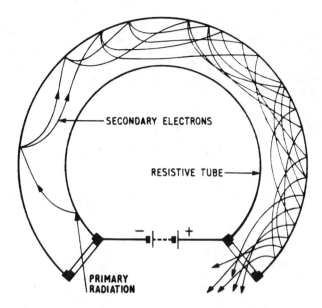

Fig. 4.15 Single channel electron multiplier constructed in a
circular configuration to reduce the ionic feedback.

Application of Photomultiplier Tubes

Due to the very large internal amplification of the
photomultiplier it is used extensively for the detection of very low
photon fluxes, down to the level of a few photons per second. To
enable these low intensities to be measured the photomultiplier is
often used in conjunction with a pulse counter, which is triggered by
the arrival of each electron pulse generated in the tube (Morton
1968), or by modulating the radiation with a mechanical chopper and
synchronously detecting it using a phase sensitive detector, (Carr et
al 1966, Ridgway 1968, Nakumura and Schwarz 1968). For these
applications it is necessary to minimise the dark current by cooling
the photomultiplier.

A second use is for the detection of short pulses of radiation,
for example in scintillation counting experiments in nuclear physics
(Birks 1964). In these experiments it is necessary to minimise the
transit spread time of the photoelectrons reaching the anode, which
can be achieved by careful focussing of the electrons. Alternatively,
transmission dynodes can be employed, which require high electric
fields for maximum efficiency, which will also reduce the transit
spread time. These tubes are able to respond to microwave
frequencies, and so can be used for heterodyne detection.

Photomultiplier tubes are also used for the detection of

ultraviolet radiation, down to about 0.01μm. The tubes are often
required to be insensitive to visible radiation, (sun blind), which
can be achieved by careful choice of the photocathode and window
material.   The photoemissive materials used for these devices are
rubidium caesium telluride, caesium iodide or metal cathodes (Cairns
and Samson 1966).

4.6 Imaging Tubes

A very important application of photoemissive detectors is in
imaging tubes which convert an optical image into a corresponding
electron image, which can then either be re-displayed as an optical
image or used to produce a charge image (Nudelman 1978, Woodhead
1980).

The image which can either by visible, ultraviolet or infrared is
focussed onto a semitransparent photocathode, sensitive in the correct
spectral region, and photoelectrons are emitted with a spatial

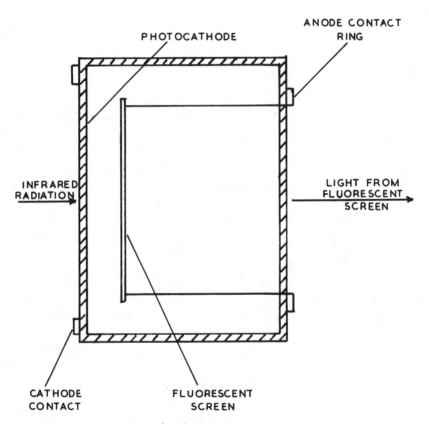

Fig. 4.16 Proximity tube.

intensity distribution which matches the focussed image. In image
intensifier and converter tubes the electrons are then accelerated
towards a phosphor screen where they reproduce the original image with
enhanced intensity or a wavelength conversion into the visible.

The simplest imaging device is the proximity tube in which the
phosphor screen is placed approximately 0.5cm from the photocathode as
illustrated in Figure 4.16, and a high electric field, typically 5kV
applied. Thus the spread of electrons emitted from the cathode is
limited and the image quality preserved. An improvement on this
device incorporates electrostatic focussing and fibre optic plates.
The image is focussed onto a fibre optic window and transmitted onto
the photocathode. The photogenerated electrons are then accelerated
onto an aluminised phosphor screen, and an inverted intensified photon
image generated, which is transmitted through the output window to
appear as a direct view image. Such an intensifier tube is shown in
Figure 4.17. The maximum gain obtained from this type of device is
approximately 2000, which limits their use in very low intensity
scenes, such as starlight.

One way of increasing the gain is by coupling several
electrostatically focussed tubes in cascade using fibre optic links
between each stage. These are known as cascade tubes, and three
stage devices are now commonly available with brightness gains of
approximately 50,000. An example of one of these tubes is shown in
Figure 4.18. A 45kV power supply is required to run this tube, and
this is often derived from an oscillator encapsulated with the
intensifier, operating from a low voltage d.c. supply such as alkaline
batteries.

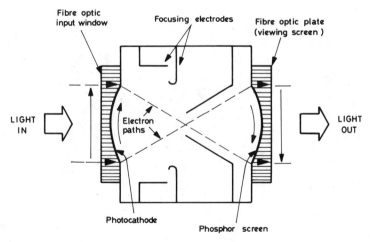

Fig. 4.17 Electrostatically-focussed single stage image intensifier.

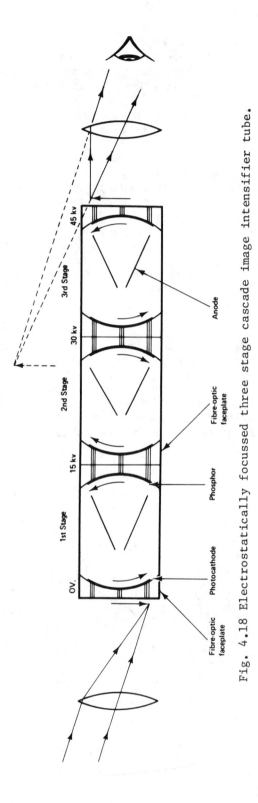

Fig. 4.18 Electrostatically focussed three stage cascade image intensifier tube.

An alternative method of increasing the gain of a single stage device is to incorporate an electron multiplier plate (as described in section 4.5.2) and gains of up to 10^5 have been obtained. The channel plate is mounted just in front of the phosphor and the photoelectrons are amplified before being finally accelerated across a small gap to the screen, as shown in Figure 4.19. To ensure that electrons travelling parellel to the axis of the tube impinge with the channel walls, the microchannel tubes are mounted at a small angle to the optical axis of the system. The spatial resolution of these tubes is limited by the lateral spread of the electrons entering and leaving the channel plate. A photograph taken using a channel plate inverter image intensifier is shown in Figure 4.20.

Image converter and intensifier tubes are often combined in a single tube for imaging of infrared scenes, when for covert operation additional visible radiation cannot be added to the scene. Although for the observation of hot objects, solid state photon detectors are now becoming more popular due to their higher sensitivity.

Image intensifier tubes are used extensively in nuclear physics experiments, for example detecting the tracks of nuclear particles passing through a scintillator (Ruggles and Slark 1964), and in astronomical observations where low dark currents and high resolution are required, and their adoption greatly reduces the exposure times needed.

Alternatively the emitted photoelectrons can be focussed onto an insulating target, to form a charge image which can be read out by scanning an electron beam across the image, and the electrical output can then be transmitted.

This forms the basis of a television camera tube, although solid state devices are now replacing them.

The colour response of the system is determined by the photocathode, and to obtain a close match to the response of the human eye the S-10 cathode has been used in many systems. Improved quantum efficiency can now be obtained using the oxidized bialkali cathode ($K_2CsSb(O)$) which also has good panchromatic response, and the gallium arsenide NEA photocathodes which have produced tubes with cathode responses of $950\mu A/lm$ (Ceckowski et al 1981).

In conclusion it can be seen that in a period of approximately fifty years from 1930, a selection of photoemissive materials suitable for many applications and operating over a range of wavelengths, has been developed. An interesting hypothesis suggested by Sommer (1983), is that all the conventional photocathodes resulted from lucky accidents, and it was not until the development of the negative electron affinity devices, that scientific innovation was first applied.

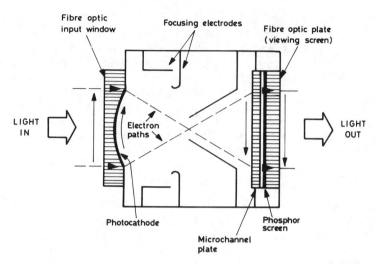

Fig. 4.19 Inverting michrochannel plate image intensifier.

Fig. 4.20 Photograph taken with a channel plate imager intensifer
under starlight conditions.

CHAPTER 5

SOLID STATE PHOTON DETECTORS

5.1 Introduction

A second category of photon detector is the solid state internal detector in which the incident photon does not cause the emission of an electron from the surface of the material as happens for photo-emissive devices, but the photoexcited carrier remains within the sample.

The principle of operation for all internal photon detectors is that incident radiation causes the excitation of bound carriers in the material into mobile states, and this transition can be monitored as an electrical output signal. There are several different methods by which this change can be observed depending on the device structure, and some of the more common devices are listed in Table 5.1. The most widely employed methods measure the change in resistivity of a photoconductive element, or the voltage generated across a junction, and the most popular detector types and some typical performance data are discussed in the following sections.

5.2 Photoconductors

Photoconductive detectors can be divided into three main categories; intrinsic, extrinsic and free carrier devices. In all three a change in the carrier conductivity is measured, however the significant difference is the state from which the bound charge is excited.

In general a pair of ohmic contacts are fixed to the sample, and a current passed through the element, as shown in Figure 5.1(a). When radiation falls on the detector the photosignal is measured either as

Table 5.1   Classification of Photon Detectors

Detector Types			Examples
Photoconductors:	i)	intrinsic	lead salts, cadmium mercury telluride
	ii)	extrinsic	doped Ge, Si
	iii)	free carrier	InSb
Junction Devices:	i)	homojunctions	InSb, CMT, lead tin telluride
	ii)	heterojunctions	PbTe/LTT, GaAs/ $Ga_{1-x}Al_xAs$ $Al As$
	iii)	Schottky Barrier	Pt/Si
	iv)	Avalanche	Si, Ge
MIS Devices			InSb
Photoelectromagnetic (PEM)			InSb

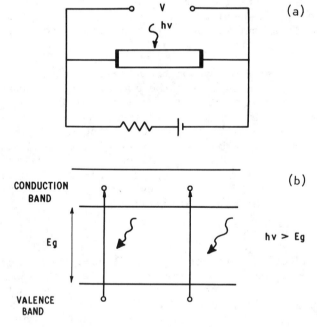

Fig. 5.1 (a) Schematic representation of a photoconductive
         element. (b) Optical excitation process in an intrinsic
         photoconductor.

a change in voltage across the load resistor in series or as a change in current through the sample.

The operation of an intrinsic photoconductor is shown in Figure 5.1(b). It requires the incident photon to have an energy $h\nu$, greater than the energy gap E_G of the material for electron-hole production; hence

$$h\nu \geqslant E_G \qquad\qquad (5.1)$$

Consequently there will always be a long wavelength threshold, beyond which no photoconductive response is obtained, and this is given by

$$\lambda = \frac{hc}{E_G} \qquad\qquad (5.2)$$

Intrinsic devices have relatively large optical absorption coefficients, particularly those employing direct energy gap materials, and also due to their relatively long excess carrier lifetimes they can generally be operated at higher temperatures than extrinsic devices.

A range of semiconductor materials has been employed as intrinsic photodectors, operating at various temperatures and a summary of the more common materials is given in Figure 5.2, with their wavelength cut-offs. It can be seen that a large number of materials are available for detection at wavelengths of less than approximately 8 μm, but beyond that no simple semiconductor compounds exist with energy gaps of 0.1 eV.

One method of extending the wavelength range has been by the use of extrinsic photoconductors. In these devices the incident photon causes the excitation of a carrier from a bound impurity state, sitting in the forbidden energy gap, to a free conduction state. This may either be a donor level sitting close to the conduction band or an acceptor level close to the valence band, as illustrated in Figure 5.3(a). The long wavelength limit will now be determined by

$$\lambda = \frac{hc}{E_i} \qquad\qquad (5.3)$$

where E_i is the impurity ionization energy.

Thus by careful choice of an impurity dopant, detectors have been fabricated which operate out to wavelengths of several hundred microns.

The major disadvantage of extrinsic devices is the need to operate them at much lower temperatures than instrinsic devices sensitive at

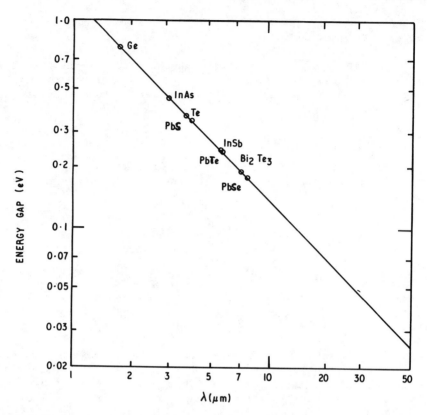

Fig. 5.2 The energy gap versus cut-off wavelength, and a range of
intrinsic photodetector materials of use in the infra-red
region.

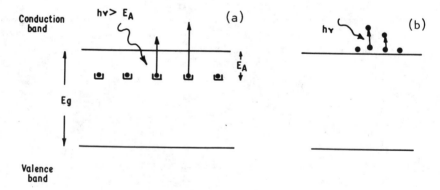

Fig. 5.3 The optical excitation process for (a) an extrinsic
photoconductor and (b) a free carrier photoconductor.

the same wavelength. Typical operating temperatures are in the 4-30K
range. This is partly due to the much lower optical absorption
coefficient of extrinsic devices, which requires the detectors to be
thicker and partly to the very high captive cross section ($\sim 10^{-12} cm^2$)
of the ionized centres, for free electrons. The latter means that
when even a small fraction of the impurity centres are thermally
ionized very short lifetimes of the photo-generated carriers can
result.

To achieve detection beyond approximately 300 µm, a third
photoconductive mechanism is employed, free carrier conduction. These
detectors have been fabricated from semiconductors with very high
carrier mobilities, such as indium antimonide (Putley 1960, 1961,
1964[2]). The incident radiation causes intraband transitions of the
electrons in the conduction band, changing the electron mobility and
hence the resistance of the material, as shown in Figure 5.3(b). It
is necessary to cool the devices to liquid helium temperatures to
reduce the rate of transfer of radiative energy to the crystal
lattice, and to produce a significant change in the electron
distribution in the conduction band. These devices are used mainly at
long wavelengths where the free carrier absorption is high, typically
in the 1000-2000 µm region. At shorter wavelengths the absorption
coefficient falls rapidly as the square of the wavelength,
(Shivanandan et al 1975).

If a photoconductive element is placed in a transverse magnetic
field and incident radiation generates electron-hole pairs by
instrinsic absorption, then as the carriers diffuse through the sample
they will be deflected by the magnetic field; this effect is known as
the photoelectro-magnetic effect, PEM. The electrons and holes will
be deflected in opposite directions through the material, thus
establishing a longitudinal electric field, which can be detected
either as an open circuit voltage or a short circuit current, (Kurnick
and Zitter 1956).

The PEM effect is often used as a method of determining the
surface recombination velocity of materials, and has only found
limited use as an infrared detector as it offers very little advantage
over conventional photoconductive or photovoltaic devices and also
requires a magnet. One application has been for low sensitivity
devices operating at room temperature, where the high mobility
material InSb, has sufficient sensitivity at wavelengths shorter than
about 7 µm, to enable its use for certain industrial applications,
(Kruse 1959).

5.3 Junction Detectors

The junction or photovoltaic detector consists of two regions of
material with an internal potential barrier which produces a depletion
layer having current rectifying properties. Most photovoltaic devices

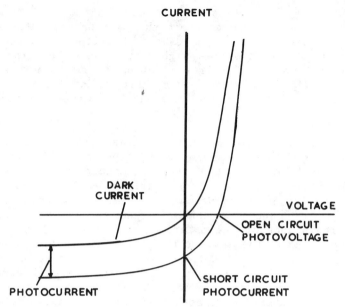

Fig. 5.4 The current-voltage characteristic of a photovoltaic
        detector.

employ instrinsic absorption, although it is possible to produce an
extrinsic photovoltaic effect (Ryvkin 1964).

The common features of all photovoltaic devices are that the
incident photons are absorbed in the material and create electron-hole
pairs, and the minority carriers diffuse to the junction where they
are swept through the depletion region, thus forward biasing the
junction to produce either an open circuit voltage or a short circuit
current. Consequently the current-voltage characteristic of such
devices is no longer linear, as was the case for photoconductors, but
now shows rectifying properties, as shown in Figure 5.4. With no
radiation incident upon the device a reverse bias current, called the
dark current, is observed, if the detector is illuminated with
radiation of the correct wavelength the I-V characteristic will be
displaced and the signal can be measured, this is given by

$$\Delta i \quad = \quad -\eta q A \phi \tag{5.4}$$

where A is the detector area, ϕ the incident photon flux and η the
quantum efficiency, defined as the number of optically generated
carriers crossing the junction per incident photon. The current-
voltage characteristic of a diode is given by

$$Id \; = \; Is \; [\exp(\frac{qV}{\beta kT}) - 1] \tag{5.5}$$

where Id is the measured diode current, Is is the dark reverse-bias saturation current, V is the applied voltage, k is Boltzmann's constant, β is a constant of the order of unity and T the absolute temperature.

To obtain a high quantum efficiency in these devices it is important to ensure that all the incident radiation is absorbed within a diffusion length of the junction, consequently the front surface should have a high absorption coefficient, the junction depth should be small compared to a diffusion length and the recombination velocity at the front surface should be low.

The current responsivity R, of a photodiode is derived from the photocurrent Δi such that

$$R_I \; = \; \frac{\eta q}{E_\lambda} \tag{5.6}$$

$$R_I \; = \; \frac{\eta q \lambda}{hc} \tag{5.7}$$

In the normal junction detector the device is fabricated with a p type and an n type region formed in the same material, these are known as homojunctions. However, it is possible to produce a hetero-junction structure from two different semiconductors having a similar lattice spacing and structure. Epitaxial growth is often used to produce these devices with the wider energy gap material being used as the front face window, to allow radiation at the wavelength of interest to be absorbed in the substrate, away from the front surface, thus improving the quantum efficiency.

Two of the most important heterojunction structures are GaAs - $Ga_{1-x}Al_x$ As and PbTe - $Pb_{1-x}Sn_xTe$.

Similar effects to those of a p-n junction are produced by a metal-semiconductor interface, known as a Schottky Barrier. Generally a thin metal film is deposited onto the semiconductor material and the device can be illuminated through the metal layer if it is semi-transparent in the wavelength range of interest, alternatively the detectors are back illuminated through a suitably thin semiconductor. These devices are most applicable for semiconductors in which both n and p-type materials cannot be easily prepared for homojunction devices.

The photovoltaic structures described in the previous sections have no internal amplification but large gains can be achieved with an avalanche photodiode. However, the signal to noise ratio will not be improved.

The detector is operated in reverse bias, where both thermally and photoexcited carriers are accelerated within the high field region of the junction such that they are able to excite additional carriers by collision with the lattice atoms (D'Asaro and Anderson 1966). Devices have mostly been fabricated from silicon and germanium and gains of 10^6 at 10 KHz and 10 at 10 GHz have been achieved, although indium arsenide diodes have been investigated, to extend the wavelength band (Lucovsky and Emmons 1965).

To obtain a low noise high gain avalanche device it is necessary to have a large ratio of the ionization coefficients for electrons and holes. The silicon and germanium devices operating at approximately 0.8 μm have a ratio of typically fifty, but, if the wavelength band is extended by using III − V compounds, this ratio becomes very close to unity, which causes an increase in the avalanche multiplication noise. However, it has been recently demonstrated that a new multilayer structure enables an enhanced ionization rate ratio to be obtained. These devices have intrinsically lower noise and operating voltages than conventional avalanche detectors (Williams et al 1982, Capasso 1983).

The detector consists of a graded band gap multilayer structure, in which the ionization energy is provided entirely by the hetero-interface conduction band steps. Thus, ideally only electrons will cause ionization. The band structure of such a multilayer device is shown in Figure 5.5. At each stage the composition is graded from a narrow energy gap Eg_1, to a large band gap Eg_2, with an abrupt return to the low gap material. For most III − V heterojunctions the largest discontinuity occurs in the conduction band, whilst only a small change is observed in the valence band. If an electric field is now applied the band structure will be modified as shown in Figure 5.6. Consequently if a photoelectron is generated near the p+ contact no ionization will occur in the first graded region as the energy will be insufficient, however if the conduction band discontinuity is

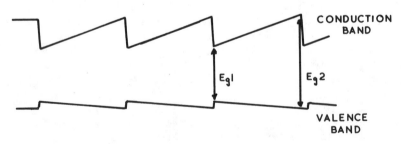

Fig. 5.5 The band diagram of an unbiased graded multilayer material.

comparable to or greater than the electron ionization in the narrow
gap material the electron will impact ionize on crossing this
junction, and the process will then be repeated down the structure,
multiplying up the carriers at each stage, in a similar manner to that
observed at the dynodes of a photomultiplier. The discontinuities in
the valence band are in the wrong direction to assist ionization, thus
the only hole ionization is due to the applied electric field.

Ideally the electron gain at each stage should be two, but it is
reduced by δ, which is the fraction of electrons which do not impact
ionize. Hence for a device of N stages the total gain will be given
by

$$M = (2-\delta)^N \qquad\qquad (5.8)$$

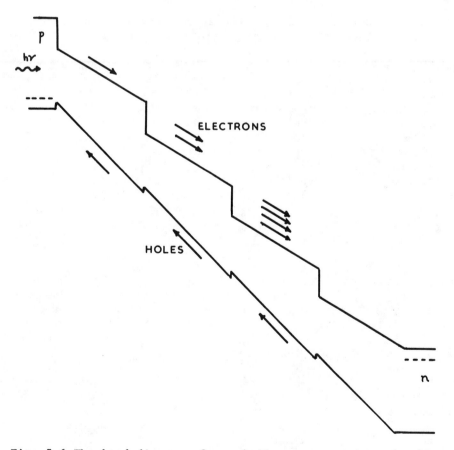

Fig. 5.6 The band diagram of a multilayer structure under bias,
showing the electron gain in the conduction band.

To minimize δ, the transition from the large to narrow gap material must be smaller than the mean free path for phonon scattering (50-100 $\overset{o}{A}$).

For operation in the 1.3-1.6 μm range two material systems have been studied, AlGaAsSb/GaSb on a GaSb subtrate and InGaAlAs/InGaAs on an InP substrate. In the AlGaAsSb system the minimum electron ionization energy is 0.80 eV (GaSb) and the maximum hetero-interface band gap difference is 0.85 eV (AlAs$_{0.8}$Sb$_{0.92}$ to GaSb) of which 0.72 eV occurs in the conduction band step. The 0.08 eV deficit is obtained from the electron drift field.

Another important advantage of these detectors is the low operating voltage required. For a five stage device, with each layer 3000 $\overset{o}{A}$ thick, to achieve a gain of approximately 32, the operating voltage is only 5 volts. This is due to the abrupt change in electron energy at the band edge discontinuities.

To produce these devices molecular beam epitaxy is the most convenient method, at growth temperatures of between 500 and 650°C.

These new devices offer potentially superior noise performance and operate at significantly lower voltages than conventional avalanche photodiodes. They also extend the sensitivity into the near infrared region of the spectrum, and can be considered to be the solid state version of the photomultiplier tubes described in section 4.5.2.

One major advantage of junction detectors is that unlike photo-conductors they require no bias supply. Consequently, there is a considerable simplification of the associated electronics, and reduced heat load for cooled devices. Photodiodes may also be directly coupled into charged coupled devices, without requiring an interfacing amplifier, and can be fabricated as planar structures for use in large two dimensional arrays.

The theoretical maximum detectivity of a photodiode is approximately 40% higher than that of a photoconductor, due to the absence of any recombination noise. To achieve a high quantum efficiency with this type of detector the junction depth should be small compared to the carrier diffusion length.

5.4 Metal-Insulator-Semiconductor Detectors

As previously discussed the modern TV camera tube now employs a silicon charge coupled device, CCD, operating in the visible region, and both linear and two dimensional arrays have been produced, (Barbe 1975). Similar types of metal-insulator-semiconductor devices are now available for detection of infrared radiation either in the form of a CCD, or a charge injection detector, CID. A typical MIS structure is shown in Figure 5.7. By applying a voltage to the

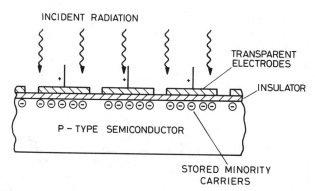

INCIDENT RADIATION

TRANSPARENT
ELECTRODES

INSULATOR

P – TYPE SEMICONDUCTOR

STORED MINORITY
CARRIERS

Fig. 5.7 Schematic representation of a metal–insulator–semiconductor
(MIS) device.

surface electrode of correct polarity, depending on the semiconductor
type, a depletion region can be created, hence when incident
radiation generates electron-hole pairs by intrinsic absorption in the
semiconductor, the minority carriers will be stored in this region,
whilst the majority carriers will be lost in the bulk.

These devices in general have high optical absorption thus
producing detectors with a high quantum efficiency.

5.5 Examples of Photon Detectors

5.5.1 Intrinsic Photoconductors

Photoconductive detectors for the visible and near infrared region
have been studied for many years, with particular interest in the
development of materials and detectors for operation at longer
wavelengths for thermal imaging applications.

In the visible spectrum photoconductive cadmium sulphide and
cadmium selenide devices are the most common. The detectors are
generally fabricated by depositing a layer of the sensor onto a
substrate followed by the bias electrodes. The peak in the sulphide
response curve occurs at approximately 0.6 µm and the selenide at
0.72 µm, which closely matches the response of the eye. Cadmium
sulphide devices are generally more stable with a lower dark current
than the selenide detectors but their response is limited to
approximately 100 Hz, which is about an order of magnitude below that
of CdSe. The devices operate at temperatures up to 60–70°C, and have
a linear response with varying input power.

A general review of their characteristics is given by Trautwein
(1964). An alternative material investigated is cadmium
sulphoselenide which exhibits intermediate properties of the sulphide
and selenide.

To extend the response into the near infrared region one of the first materials investigated in about 1920, by Case, was thallium sulphide which gave a response out to approximately 1.1 μm (Cashman 1959). But the limited wavelength response of these devices restricted their development. For applications requiring detection of 1 μm radiation it is now more convenient to use silicon or germanium photodiodes, which are sensitive to wavelengths of approximately 1.0 μm and 1.7 μm respectively (Loh 1963).

The next group of materials to be studied were the lead salts, (sulphide, selenide and telluride) which extended the wavelength response to 7 μm. Lead sulphide devices were originally fabricated in the 1930's from natural galena found in Sardinia, (Kutscher 1975). However, for any practical applications it was necessary to develop a method of producing synthetic crystals, and both vacuum evaporation and chemical deposition techniques have been adopted. The polycrystalline evaporated films have the faster response times, typically 100 μsec, but cannot produce such large area devices as the chemically deposited layers. The room temperature cut off is 2.5 μm. Lead selenide detectors were fabricated by chemical deposition and the wavelength band was extended to 3.4 μm. To improve the performance of both PbS and PbSe devices, after deposition the layers are heated in oxygen.

If the devices are cooled, a significant increase in the sensitivity and a shift in the peak of the response curves towards longer wavelengths occurs, thus PbSe will detect radiation beyond 6 μm when cooled to 77K, and PbS to approximately 3 μm. During the Second World War the Germans produced systems using PbS detectors which were able to detect hot aircraft engines against a night sky at significant distances and research was under way to produce infrared homing devices for missile applications.

Unfortunately the fabrication methods developed for these thin layers of PbS and PbSe are not completely understood, consequently the possibility of using single crystals has been studied. Kimmitt and Prior (1961) produced single crystal PbSe photoconductors, but due to the difficulties of preparing the crystals with sufficient purity they have not been developed.

The third lead salt investigated was PbTe, which has a band gap of 0.3 eV, and can be employed at wavelengths in the region of 2 to 5.5 μm, but to obtain satisfactory performance the detectors must be cooled.

The performance obtained with lead salt detectors is very good, PbS devices operating at 195K have peak D^* values above 1×10^{11} cm $Hz^{\frac{1}{2}}$ W^{-1} and by cooling to 77K values exceeding 6×10^{11}

cm Hz$^{\frac{1}{2}}$ W^{-1} have been obtained, which are comparable to any other
detectors operating at 2 μm, and PbSe devices have been produced with
detectivities of 2 x 10^{10} cm Hz$^{\frac{1}{2}}$ W^{-1} and responsivities of 6 x
10^3 VW^{-1} at 193K (Moss et al 1973, Bode 1966). However, they do have
several disadvantages. Due to their fabrication processes not being
fully understoood, uniform, stable layers are difficult to produce,
and also their impedances are rather high. More importantly, their
response times are rather long, typically greater than 100 μsec,
consequently other types of photoconductive detectors have been
developed in particular indium antimonide and indium arsenide.

These are III-V compound semiconductors, which can be operated
both at room temperature and with increased performance at lower
temperatures, which also increases the energy gap thus reducing the
long wavelength limit. InSb has a room temperature energy gap of
approximately 0.18 eV, which corresponds to a threshold wavelength of
about 7 μm, on cooling to 77K these change to 0.23 eV and 5.5 μm.
Devices operating at 77K are very close to the performance of an
ideal detector, and commercially available detectors are usually in
the range 5 x 10^{10} to 1 x 10^{11} cm Hz$^{\frac{1}{2}}$ W^{-1}, with responsivities in
excess of 10^5 VW^{-1}, and response times typically 5 μsec. As the
operating temperature is increased the Johnson noise becomes the
dominant mechanism, thus reducing the detector performance, but
response times of less than 10^{-7} sec can be achieved. The
photoconductive process in InSb has been studied extensively, and more
details can be found in Morton and King (1965), Putley (1977) and
Elliott (1981).

Indium arsenide is a similar compound to InSb, but has a larger
energy gap, so that the threshold wavelength is 3-4 μm, and both
photoconductive and photovoltaic detectors have been fabricated.

As can be seen from Figure 5.2 there are no suitable elemental
or simple compound semiconductors available to extend the wavelength
range beyond approximately 7 μm, consequently most of the devices used
at longer wavelengths were originally extrinsic detectors using
various impurities in germanium, see section 5.5.2. However, with the
development of the ternary alloy systems (Lawson et al 1959) it is now
possible to control the energy gap of the material to meet a specific
wavelength requirement.

The two most developed materials are mixed compounds of mercury
telluride and cadmium telluride, and of lead telluride and tin
telluride, and by adjusting the Hg/Cd or Pb/Sn ratios the energy gap
can be continuously varied to produce material suitable for intrinsic
detection from approximately 2 μm up to 30 μm.

In the cadmium mercury telluride, CMT system the energy gap varies
from approximately 1.6 eV (CdTe) to an effective negative energy gap
of approximately 0.2 eV for HgTe, which is a semimetal, see Figure

5.8. The energy gap in the alloy varies almost linearly with
composition throughout the range, and passes through zero at x = 0.15,
where x is the mole fraction of CdTe.   For use in the 8–13 μm
atmospheric window which corresponds to an energy gap of 0.1 eV, a
composition of x = 0.20 is required for 77K operation.

A great deal of work has been invested in the development of CMT
devices during the last twenty years, due to its attractive
fundamental properties.   The material has a low intrinsic carrier
concentration, a high mobility ratio, and long minority carrier
lifetimes associated with intrinsic recombination processes.   Borrello
et al (1977) have reported lifetimes of several microseconds for
x = 0.2 material at 77K, and Kinch et al (1977) and Elliott (1981)
have measured lifetimes greater than 10 μsec for x = 0.3 material
at 193K.

The major difficulty with the CMT system is in the growth of
single crystal material, due mainly to the very high vapour pressure

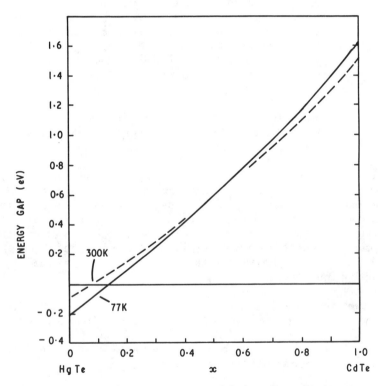

Fig. 5.8 The energy gap versus composition for the ternary alloy
         system $Hg_{1-x}Cd_xTe$ at 300K and 77K   (Schmit and Steltzer
         1969).

of mercury over the material at the melting point and the large
separation of the liquidus and solidus in the phase diagram.   Until
recently the majority of devices were fabricated from bulk grown
material using either a Bridgman method (Bartlett et al 1969) or a
solid state recrystallization technique (Long and Schmidt 1970).
Ingots, typically 15 cm long and 10–15 mm in diameter, are grown and
are then cut into thin slices which are carefully polished to remove
any damage.   To produce photoconductive elements an n-type slice is
polished to approximately 10 μm thick, and the array then delineated.
Elements, typically, 50 μm square are produced in arrays of various
shapes, and a forty eight element photoconductive array on a 62.5 μm
pitch, produced by Mullard Ltd, UK, is shown in Figure 5.9.

Recently the possibility of growing epitaxial layers of CMT have
been investigated and several techniques including liquid phase,
vapour phase, sputtering and molecular beam epitaxy have produced
successful layers.   The major advantages of an epitaxial growth method
are the possibility of growing large area layers, determined only by
the availability of good quality substrates, CdTe being the most
common substrate used, and the facility to produce heterostructures.

CMT photoconductive elements for operation in the 8–14 μm band at
77K are now widely available with high performance and in large
arrays.   The detectors are generally fabricated from n-type material
with a carrier concentration of approximately 5 x 10^{14} cm^{-3}.
Responsivities of typically 5 x 10^4 VW^{-1} and D* of 8 x 10^{10}
cm Hz$^{\frac{1}{2}}$ W^{-1} have been obtained in a 30° field of view, for a peak

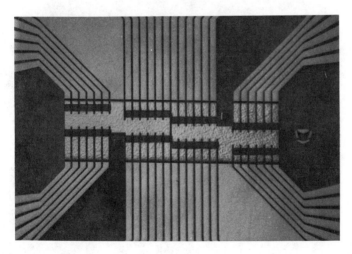

Fig. 5.9 Photograph of a 48 element Hg$_{1-x}$Cd$_x$Te array in the form of an
         8 x 6 matrix.   The sensitive area of each element is 50 μm x
         50 μm and the pitch along a row is 62.5 μm   (Courtesy of
         Mullard Ltd).

wavelength of 12 μm. There is much interest in operating these
devices at elevated temperatures on thermoelectric coolers, but the
lifetimes are much shorter. Elliott (1981) have calculated the
performance of these devices and shown that it is advantageous to
fabricate these devices from p-type material to obtain a low electron
concentration and thus reduce the Auger recombination process which
dominates. Peak detectivities of typically 10^9 cm $Hz^{\frac{1}{2}}$ W^{-1} at 10 μm
have been obtained at 193K.

In the 3-5 μm band the devices are generally operated at around
200K, where the dominant intrinsic recombination process is
radiative. The detectors produced approach the theoretical value,
with D* values of 2 x 10^{11} at 4 μm and 5 x 10^{10} cm $Hz^{\frac{1}{2}}$ W^{-1} at 5 μm.

A major advance in the last few years has been the development of
a new type of photoconductor for use in scanned thermal imaging
systems, this is known as the TED or SPRITE (signal processing in the
element) detector, (Elliott 1982, Elliott et al 1982). The important
benefit this device achieves is that the time delay and integration
required in serial scan thermal imaging systems is performed within a
single detector element. In a conventional serial scan system the
image of the scene is scanned across a series of discrete detectors,
the output for each device is then amplified and delayed by the
correct amount so that all the detector outputs can be integrated in
phase, as shown in Figure 5.10.

In the SPRITE detectors these functions are actually performed in
the element and a single device is illustrated in Figure 5.11. It

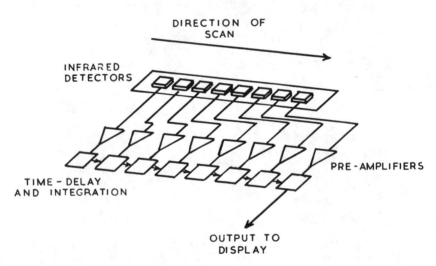

Fig. 5.10 Conventional serial scan system, with the output from each
detector amplified and time delayed and integrated.

consists of a long strip of n-type cadmium mercury telluride, typically 700 μm long, 50 μm wide and 10 μm thick, with three ohmic contacts. When radiation is incident on the semiconductor, electron-hole pairs are produced, and if a constant current bias is applied across the whole element the photogenerated carriers will move under the influence of the electric field with a velocity given by the product of the field and the ambipolar mobility. Thus the minority carriers, the holes, will move towards the cathode, now if the scene image is scanned along the element at a speed matching the drift velocity, the photosignal will be continuously increased for the duration of the minority carrier lifetime. The signal is measured at the readout contact as the charge modulates the conductivity as it moves through the readout zone.

The performance of the SPRITE detector can be specified in terms of the number of equivalent conventional elements it represents. The integration length L of the device is determined by the minority carrier lifetime τ the applied electric field E, and the ambipolar mobility μ such that

$$L \ = \ \mu E \tau \qquad\qquad (5.9)$$

The noise voltage will also be integrated over this length. The number of resolution elements which the device is equivalent to is the integration length divided by the resolution length 1, which for this device is the readout length, thus Neq is given by

$$Neq \ = \ \frac{\mu E \tau}{\ell} \qquad\qquad (5.10)$$

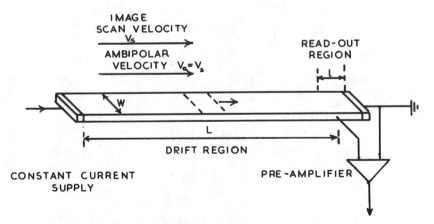

Fig. 5.11 Schematic representation of the operation of a SPRITE detector, in which the image scan velocity is matched to the ambipolar velocity.

Now $\ell/\mu E$ is the time taken for the minority carriers to drift through the readout region τ_{ro}, which is the reciprocal of the pixel rate S, that is the rate that information is handled by the system.

$$\therefore \quad Neq = \frac{\tau}{\tau_{ro}} = s\tau \tag{5.11}$$

Thus the detectivity improvement obtained will be

$$D^*(\text{SPRITE}) \backsim D^*(\text{single element}) \sqrt{s\tau} \tag{5.12}$$

Consequently to achieve significant benefit with this device it is necessary to produce long lifetime detectors, and to operate them such that the data rate is sufficiently large for the transit time to be substantially less than the minority carrier lifetime in the filament.

SPRITE detector arrays have now been fabricated and demonstrated in thermal imaging systems, operating in both the 3-5 μm band at thermoelectric temperatures and in the 8-14 μm band at 77K. The most important system developed with these devices has been the United Kingdom Thermal Imaging Common Module (UKTICM) Class II imager produced by Marconi Avionics Ltd and Rank Taylor Hobson (Cuthbertson and MacGregor 1981) which produces a TV compatible output. The detector used in this system is shown in Figure 5.12, and consists of eight parallel SPRITE detectors, and an example of the imaging achieved in this type of system is shown in Figure 5.13. Each element has a bias and a readout contact at each end which increases the yield in fabrication, although, obviously once installed in a

Fig. 5.12 Photograph of and eight-row SPRITE detector as used in
the UKTICM Class II imager (Courtesy of Mullard Ltd).

system the detector is only operated in one direction. The device is typically 700 μm long, 62.5 μm wide with a 62.5 μm long readout region. The blackbody D^* values obtained are of the order of 1.1 x 10^{11} cm $Hz^{\frac{1}{2}}$ W^{-1} in a f/2.5 field of view and a responsivity of 6 x 10^4 VW^{-1}. Consequently the eight element array is equivalent to between 100 and 200 conventional photoconductive elements without any of the complexities of time delay and integrate circuitry. Arrays with increased parallel and serial content have also been fabricated.

5.5.2 Extrinsic Detectors

Before the development of the ternary alloy systems the only method of extending the wavelength of operation of photoconductors beyond approximately 8 μm was by using extrinsic devices. The most widely used materials for extrinsic photoconductive detectors have been germanium (Emmons 1970) and silicon (Long 1967[2], Soreff 1967) and high performance devices have been produced operating from 0.5 to 150 μm with responsivities of up to 10^4 VW^{-1} and detectivities of 2 x 10^{10} cm $Hz^{\frac{1}{2}}$ W^{-1}.

Germanium has been doped with a wide range of impurities including

Fig. 5.13 Photograph of thermal image obtained with the UKTICM Class II system (Courtesy of Marconi Avionics and Rank Taylor Hobson).

Continued overleaf

Fig. 5.13 Photographs of thermal images obtained with the UKTICM Class
         II system (Courtesy of Marconi Avionics and Rank Taylor
         Hobson).

gold, mercury, copper, zinc, boron and gallium. However, as the
wavelength range is extended, the impurity level becomes shallower and
the detector operating temperature must be lowered. Gold produces a
peak sensitivity at 5 μm at 77K, mercury extends the wavelength to
14 μm at 30K and with boron and gallium the response will extend to
approximately 120 μm but the detector must be cooled to 4K (Putley
1966, 1969). A summary of the properties of the most important
germanium doped detectors is shown in Table 5.2.

Due to the extra cooling required for extrinsic detectors, their
use has been limited to a few specialized applications. However,
recently there has been a renewed interest in the fabrication of doped
silicon detector arrays, which permits the production of a monolithic
package which incorporates both the detector and any sophisticated
signal processing which is now available. Thus offering the
possibility of low cost but complex detector-processing packages.

The instrinsic response of pure silicon is limited to
approximately 1 μm, but by the introduction of impurities this has
been extended well into the infrared region. Several different
dopants have been investigated and the response extended to 24 μm
with arsenic. These devices require cooling to temperatures in the
liquid helium range, and most effort has been directed towards finding
suitable dopants for operation in the atmospheric windows, at 3-5 μm
and 8-14 μm. Bryan (1983) has investigated the parameters controlling

Table 5.2   Extrinsic Detectors

Semiconductor	Dopant	Energy Gap Ei (eV)	Cut Off Wavelength (μm)	Temp. (K)
Ge	Au	0.15	8.3	77
	Hg	0.09	14	30
	Cu	0.041	30	15
	Zn	0.033	38	4
	B	0.010	120	4
Si	Tl	0.23	4.3	78
	S	0.17	6.8	78
	In	0.15	7.4	78
	Mg	0.11	12.1	5
	Ga	0.074	17.8	27
	Bi	0.069	18.7	29
	Al	0.067	18.4	27

the operating temperature of extrinsic silicon devices, and highlights
the dependence on the capture cross section.

Sclar (1976) reports the use of aluminium, gallium, bismuth and
magnesium for 8-14 μm devices, and indium, sulphur and thallium in the
3-5 μm band. The long wavelength devices required cooling to between
5 and 30K, and the 3-5 μm ones to 78K, and spectral response curves
were obtained. The properties for the various dopants are summarised
in Table 5.2.

The peak detectivities obtained for the 8-14 μm devices were in
the range 2 to 7 x 10^{10} cm $Hz^{\frac{1}{2}}$ W^{-1}, corresponding to quantum
efficiencies of between 3 and 35%. The 3-5 μm detectors had
detectivities of up to 1 x 10^{11} cm $Hz^{\frac{1}{2}}$ W^{-1}, and quantum efficiencies
of 48%.

For detection beyond that achieved with extrinsic germanium,
Stillman et al (1971, 1977) have investigated doped gallium arsenide
devices. They report detectors operating between 100 and 350 μm,
although the donor ionization energy corresponds to a cut-off
wavelength of approximately 214 μm. It is proposed that the longer
wavelength response is due to the photoexcitation of electrons to
excited impurity states, with subsequent further thermal excitation
into the conduction band. These sensors are usually operated at
liquid helium temperatures, 4K, as the response falls rapidly as the
temperature is raised.

5.5.3 Junction Detectors

There are very few photodiodes which are sensitive in the ultraviolet region although the response of selenium cells extends down to
approximately 0.25 μm, and there are reports of silver sulphide cells
sensitive at 0.22 μm (Galavanova and Zlatkin 1965) and Weissler (1965)
has presented results of a silicon photodiode which responded down to
0.058 μm. Silicon diodes are also used for detection through the
visible and near infrared region out to 1.2 μm with a peak at 0.9 μm,
and peak detectivities of 2.5 x 10^{12} cm $Hz^{\frac{1}{2}}$ W^{-1} have been achieved.
The response of germanium photodiodes peaks at 1.5 μm and are sensitive
out to 1.8 μm, generally their responsivities are higher than those
obtained from silicon devices but they are slower and their maximum
operating temperature is lower, typically 50°C compared to 125°C. D^{*}
(peak) values of 4 x 10^{11} cm $Hz^{\frac{1}{2}}$ W^{-1} have been reported.

Gallium arsenide, phosphide and selenide have also been studied
with responses over similar spectral ranges, and efficient gallium
arsenide solar cells have been fabricated.

The most important photovoltaic devices developed for use in the
near infrared region are the lead salts, indium arsenide and indium
antimonide. Lead salt detectors have been produced by vapour
deposition onto barium fluoride, followed by the deposition of a lead

metallization layer. The exact method of operation of these devices
is not fully understood, as the junction could either be caused by
metal diffusion into the lead salt or the production of a Schottky
barrier.

Hohnke and Holloway (1974) have studied PbSe detectors produced in
this way on BaF_2 substrates and found the devices to be background
limited down to the Johnson noise limit, with peak D^* values of 5 x
10^{11} cm $Hz^{\frac{1}{2}}$ W^{-1} when operated at 77K, and a peak response at 6.1 μm.
The quantum efficiency obtained was between 60 and 70%.

Indium arsenide and antimonide photovoltaic detectors have been
available for many years and are generaly fabricated by impurity
diffusion or ion implantation. InSb devices are operated at 77K and
have a peak sensitivity between 5.0 and 5.5 μm. Slawek (1969)
reports results from impurity diffused detectors, with peak
detectivities of 7.0 x 10^{11} cm $Hz^{\frac{1}{2}}$ W^{-1} in a 15° FOV. Foyt et al
(1970) have produced n-type layers on p-type InSb by proton
bombardment, which is the reverse of most diffused diodes. The zero
bias resistances of the detectors are typically a few hundred kilohms,
for a twenty micron diameter device. The peak detectivities in a 2π
FOV were 1 x 10^{11} cm $Hz^{\frac{1}{2}}$ W^{-1} with quantum efficiencies of about 35%.

Ion implantation of sulphur and zinc into InAs and InSb
respectively has produced p-n junctions (McNally 1970). InAs devices
have a peak sensitivity at approximately 3.1 μm when operated at 77K,
peak detectivities of 1.3 x 10^{11} cm $Hz^{\frac{1}{2}}$ W^{-1} in a 2π FOV were
obtained. The detector arrays had a uniformity of four per cent of
the mean D^* value. The InSb detectors had D^* (2π) values of 7 x
10^{10} cm $Hz^{\frac{1}{2}}$ W^{-1}.

To extend the wavelength of operation the ternary alloy systems
have again been developed. However, due to the extremely high
performance obtained from photoconductive CMT detectors there has been
much less research aimed at producing photodiodes in this material.
But with the recent interest in large staring arrays, the importance
of fabricating CMT photovoltaic devices has rapidly increased.

CMT photodiodes have been reviewed in detail by Long and Schmit
(1970) and Reine et al (1981), and junctions have been produced by ion
implantation, impurity diffusion and both in and out diffusion of
mercury. The most versatile method used has been ion implantation,
which allows precise control of the doping profile, and in particular
the possibility of producing very shallow junctions. Most of the work
has concentrated on producing n on p junction devices, using a
variety of ions, protons and electrons. The most probable reason for
the greater success in converting p-type material is that any damage
introduced by the ion beam is n-type, thus assisting the production
of the junction, hence many of the devices fabricated in this manner
may be due to damage rather than an electrically active implanted

however the highest values for the long wavelength region have been
reported by Ameurlaine et al (1973) and are in excess of 1×10^{11}
cm $Hz^{\frac{1}{2}}$ W^{-1}.

The role of various impurities when diffused into CMT has been
studied by Johnson and Schmit (1977), but the only two dopants studied
in detail for device fabrication have been indium (Scott and Kloch
1973) and gold (Kohn and Schlickman 1969).

Both heterojunction and homojunction devices have been fabricated
in polycrystalline CMT, from layers produced by triode sputtering from
a ternary alloy target, in a mercury plasma system, (Cohen-Solal et al
1974, 1976). Films of thickness between 1 μm and 5 μm, were deposited
onto hot substrates of either CdTe or CMT, and they could be doped
both p and n-type by co-sputtering from either a gold or an aluminium
target. Diodes operating in the 2 to 14 μm spectral range have been
produced with peak detectivities of 4×10^{10} cm $Hz^{\frac{1}{2}}$ W^{-1} and a quantum
efficiency of 60%, at 10.6 μm, which is comparable with the results
obtained by ion implantation.

More recently CMT photodiodes have been produced by successive
growth of p and n type liquid phase epitaxial layers on CdTe
substrates (Shin et al 1980). The p-type layer was obtained as grown,
without any intentional doping, and the n^+ layer was doped with
indium. The devices had a cut-off wavelength of 4.4 μm at 77K and a
peak quantum efficiency of 73%.

Schottky barrier devices have recently been produced by depositing
aluminium or chromium onto p-type CMT, (Polla and Sood 1980).

During the early development of CMT devices, due to the apparent
growth problems, other ternary alloy compounds were investigated.
The most viable alternative developed was lead tin telluride,
$Pb_{1-x}Sn_xTe$ (LTT) which now provides a second source of 8-14 μm
devices. The variation of energy gap with composition in LTT is
different from the CMT system as the band structure of SnTe is
effectively inverted with respect to that of PbTe, thus as the
composition is varied the energy gap will decrease, pass through zero
at some composition and then increase again, as shown in Figure 5.14
(Moss et al 1973). It can be seen that it is possible to fabricate an
LTT detector with a specific energy gap from two different
compositions, although in practice the higher PbTe composition
material is always used. The total energy gap range achieveable with
this system is much less than that of the CMT alloy, consequently
precise control of the compositional variation for uniform wavelength
cut-off in detector arrays is less stringent.

LTT is also an easier material to produce than CMT, it can be
grown in both bulk, using Bridgman (Calawa et al 1968) or horizontal
zone melting (Harman 1973) techniques, or various epitaxial methods

species.

The first devices reported were fabricated by proton bombardment (Foyt et al 1971), and peak detectivities of 9×10^{11} cm $Hz^{\frac{1}{2}}$ W^{-1} were obtained from 4 μm material and measured in an f/6 field of view. Fiorito et al (1973, 1975, 1978) implanted mercury to produce n on p CMT photodiodes. The implantation was carried out at room temperature and an energy of 30 KeV, with a total dose of between 10^{12} and 10^{13} ions cm^{-2}. Following implantation no thermal annealing was carried out. The peak detectivities obtained were 2.8×10^{11}, 5.9×10^{10} and 4×10^{10} cm $Hz^{\frac{1}{2}}$ W^{-1} at 77K for cut-off wavelengths of 3.7 μm, 8 μm and 10.1 μm respectively. Quantum efficiencies of typically 60% and in several cases exceeding 90% have been achieved.

Several other implant species have been investigated. Marine and Motte (1973) used aluminium at 250 KeV with a total dose of 5×10^{15} ions cm^{-2}, and a post implant anneal at 300°C for one hour. The devices at 77K had a peak detectivity of 7.3×10^{10} at 10.6 μm in a 30° FOV which corresponds to a quantum efficiency of 57%. Igras et al (1977) produced n on p junctions with Al, In, Hg and Zn which are expected to act as donors in CMT, however when using the acceptors P, N and Au they were unable to type convert n-type material. However, as no post implant anneals were performed, the results were probably dominated by damage mechanisms. Subsequently Bubalac et al (1980) have confirmed the results with donor implants, but they have also produced junctions in p-type material with lithium, which has been shown to be an acceptor when diffused (Johnson and Schmit 1977), magnesium and beryllium. None of the samples were subjected to any post implant anneals. By profiling the material they found that the junction formation mechanism is determined primarily by radiation induced defects which propogate deep into the material during the implantation process, rather than by the activation of the implant species.

To produce p on n junctions by acceptor ion implantation careful control of the post implant annealing conditions is required. Dennis (1978) produced devices using gold and subsequent annealing at 160°C. The detectors had peak detectivities of 1.2×10^{11} cm $Hz^{\frac{1}{2}}$ W^{-1} and a cut-off wavelength of 3.3 μm at 77K.

The original method of producing junctions in CMT was by the diffusion of mercury into p-type material, which fills the metal vacancy acceptor sites. The first devices were reported by Verie and Granger in 1965, for short wavelength materials; subsequently the wavelength range was extended to beyond 17.5 μm (Verie and Ayas 1967), and peak detectivities in the range 1 to 5×10^{9} cm $Hz^{\frac{1}{2}}$ W^{-1} were obtained.

More recently detectivities of 10^{10} for 8-14 μm and 10^{11} for 3-5 μm devices have been achieved (Becla and Pawlikowski 1976),

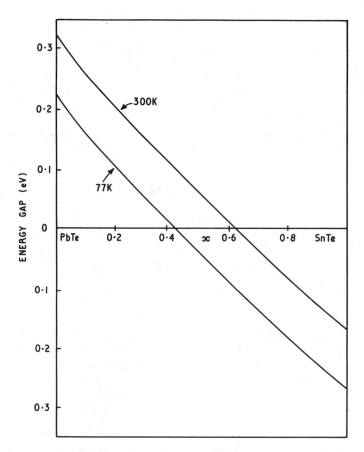

Fig. 5.14 The energy gap versus composition for the ternary alloy
        system $Pb_{1-x}Sn_xTe$ at 300K and 77K (Strauss 1968).

including liquid phase (Groves 1977) and open and closed tube vapour
phase epitaxy (Bellevance and Johnson 1976, Melingailis and Harman
1970).

        However, there are serious fundamental difficulties with LTT which
limit its applications. It is able to exist with very large
deviations from stoichiometry and it is thus difficult to prepare
material with carrier concentrations below approximately 1×10^{17}
cm^{-3}, consequently the devices fabricated are generally photovoltaic
rather than photoconductive. The second problem arises due to the
very high dielectric constant, typically 500, which limits the high
frequency response of any diodes, due to amplifier matching
difficulties.

The first photovoltaic devices produced were due to deviations in stoichiometry, introduced by isothermal annealing of p-type substrates in equilibrium with a metal rich LTT source, at temperatures in the range 400-500°C. However, although high performance devices could be produced, there were difficulties in obtaining reproducible results.

In 1975 Chia et al described results from planar arrays of homojunctions, produced by indium diffusion into p-type substrates. High, uniform quantum efficiencies were obtained with peak detectivities of 1.1×10^{11} cm Hz$^{\frac{1}{2}}$ W^{-1} measured at 80K in an f/5 field of view. Wang and Lorenzo (1977) have also fabricated high quality devices by impurity diffusion into LPE layers, with peak D* values of 2.6×10^{10} cm Hz$^{\frac{1}{2}}$ W^{-1} measured in a 180° field of view.

An alternative technology adopted for the production of long wavelength detectors is the use of heterojunctions of n-type PbTe deposited onto p-type $Pb_{1-x}Sn_xTe$ substrates, by either vapour or liquid phase epitaxy. Uniform arrays with typical peak detectivities of 7×10^{10} cm Hz$^{\frac{1}{2}}$ W^{-1} at 82K have been fabricated (Chia et al 1975).

Alternatively, for operation at longer wavelengths, lead tin selenide detectors have been produced, again by vacuum deposition onto barium fluoride substrates, and similarly by controlling the composition the cut-off wavelength can be selected. At 10 μm and 77K operating temperature, detectivities of 5×10^{10} cm Hz$^{\frac{1}{2}}$ W^{-1} have been obtained in a 26° FOV. These values are far in excess of those obtained from bulk detectors, in which the best detectivities were typically 3×10^9 cm Hz$^{\frac{1}{2}}$ W^{-1} (Rolls and Eddolls 1973).

Schottky barrier detectors operating in the ultra-violet, visible and the infrared region have been produced, and due to their high frequency response they are often used for frequency mixing applications. Gallium arsenide devices working over a large wavelength region, 50-1000 μm, have been fabricated (Hodges and McColl 1977, McColl et al 1977). In the infrared region there has been much interest recently in the possibility of constructing large detector arrays of silicon Schottky barriers, and incorporating some signal processing in the silicon, (Kohn et al 1975). The most developed devices have been fabricated using thin platinum or palladium films, which operate up to 5 μm and 3.5 μm respectively. The layers are deposited onto p-type silicon substrates and then sintered. A silicon dioxide dielectric layer is deposited, followed by an aluminium reflector, which improves the coupling of the infrared radiation to the silicide layer. This process is compatible with standard silicon integrated circuit fabrication thus allowing a monolithic detector-processor package to be developed.

The operation of the device is illustrated in Figure 5.15. The radiation is incident on the optically transparent back of the element, and photogenerated holes then cross the Schottky barrier formed between the silicide layer and the p-type silicon substrate,

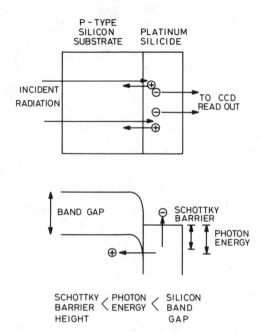

Fig. 5.15 Schematic representation of a platinum silicide Schottky
         barrier device, and the associated band gap diagram.

and the electrons are injected into the CCD read-out of the device.

   Large two dimensional arrays of these devices have now been
constructed, up to 64 x 128 (Kosonocky et al 1982) and good quality
thermal imaging demonstrated. The major advantages are that the
technology is compatible with standard silicon processing, offering
potential large area cheap focal plane arrays, with high uniformity.
However, the devices require cooling to 80K to reduce the dark
current, and also the quantum efficiency is low. The responsivity
falls rapidly with increasing wavelength, and quantum efficiencies of
5%, 1% and 0.1% have been measured at wavelengths of 3, 4 and 5 μm.
This must be compared with CMT operating at 5 μm which exhibits very
high quantum efficiencies, typically 70% and the possibility of
operation at thermoelectric temperatures. Also, due to the electrode
arrangements, requiring the incorporation of signal transfer gates
and guard rings for each detector, only approximately 25% of the
detector area is sensitive, compared to 80-90% for other systems.

   5.6 Staring Arrays

   One of the most significant recent advances in infrared technology
has been the development of large two dimensional focal plane arrays
for use in non-scanned systems, the new generation of staring arrays.

For several years large dense focal plane arrays of visible charge coupled devices (CCD), typically a few thousand elements in size, have been available for use in cameras, but the extension of this technology into the infrared region is a very new capability (Longo et al 1978).

The major advantages of a staring array system are the elimination of mechanical scanners, increased sensitivity due to the longer integration times available and the absence of scan inefficiencies. Unfortunately, the detector requirements are much more severe as high element to element response uniformity or sophisticated compensation electronics are required, and to produce sufficiently large fields of view with the required spatial resolution large numbers of elements are needed, it is also necessary to be able to handle the very large offset signals arising from the background radiation with the long integration times employed.

The early developments closely resembled the visible CCD systems and the technology is based on an MOS structure in which the charge carriers produced by the detector are fed via an input gate, into a charge storage well, where they accumulate during the stare time. The charge is then transferred by clocking the voltages on the gates, to the edge of the array where it is read out. However, the signal can only be integrated until saturation occurs in the storage well. This occurs more rapidly in the 8-14 μm band than in the 3-5 μm band, and is also more severe as the background temperature increases. With the availability of CCDs on the focal plane it is possible to incorporate additional signal processing such as background subtraction, multiplexing and time delay and integration (TDI) if scanned focal planes are envisaged.

The detector requirements for large arrays make photovoltaic devices advantageous for almost all staring array applications, they require no bias current, thus simplifying the supply circuits and eliminating the detector power loading of a cooled focal plane, they are also high impedance devices giving high injection efficiencies into a CCD. The technologies can be divided into monolithic and hybrid systems. In the monolithic approach the detector and signal processing (CCD) are both fabricated in the same piece of material, thus eliminating the need for a large number of interconnects between the detector array and the signal processing chip, which is required for the hybrid approach in which the two chips are fabricated separately and then joined, as shown in Figure 5.16. This hybrid approach allows individual testing of the two components and permits the infrared elements to be fabricated in one material and the processor in another, generally silicon

Monolithic extrinsic silicon arrays have been produced based on a standard technology, using indium and gallium doped silicon, however these arrays require operation at low temperatures, less than 40K, and

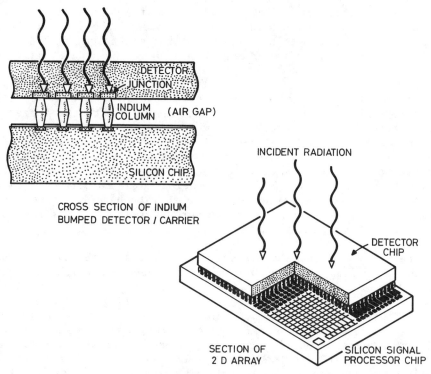

Fig. 5.16 Two dimensional hybrid structure used for infrared staring
         arrays. The cross-section shows the indium contacts used as
         electrical contact between the detector and the processing
         chip.

have relatively low responsivity and poor uniformity. Arens et al
(1983) have described the use of a 32 x 32 array of bismuth doped
silicon charge injection devices for astronomical observations.
However, the system is operated at 10K. A much more attractive
silicon system is the monolithic platinum/silicon Schottky barrier
devices described in section 5.5.3. These devices have a high
uniformity and can be operated at 80K, however they will only operate
in the 3-5 μm band, have a low quantum efficiency and at present have
a large dead area on the focal plane. However, good quality imaging
has been demonstrated, and arrays up to 64 x 128 fabricated
(Kosonocky et al 1982).

    Platinum silicide Schottky barrier devices are now being used for
astronomical applications. Ewing et al (1983) have described
observations obtained in the 1.1 μm to 3 μm wavelength range using a
32 x 64 array, mounted in a telescope.

    A monolithic cadmium mercury telluride technology has been

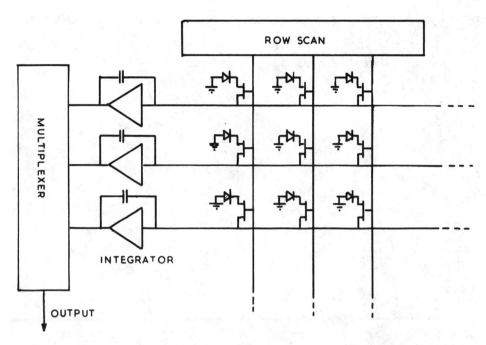

Fig. 5.17 Schematic of a line addressed array using MOS switches
         and integrators.

demonstrated, by Texas Instruments, and arrays up to 64 x 64 using
CIDs (Charge Injection Devices) have been fabricated for the 3-5 μm
and 8-10 μm band (Roberts 1983).   In this device both the storage and
readout cells are MIS devices.   At present the possibility of
extending the technology to longer wavelengths is being investigated.
The main difficulty arises due to the well capacities being limited by
band to band tunnelling in the narrower gap material, and consequently
the long wavelength cut-off of these devices has been limited to
9.2 μm, at present.

     The hybrid approach has been used to produce several different
detector/processor combinations, but two of the most important are the
InSb and CMT structures.   Back illuminated, indium antimonide arrays
have been fabricated, and hybridized with a silicon CCD processor
using an indium bump interconnect.   The device is limited to operation
in the 3-5 μm band, but as it is illuminated from the back surface
there is no obscuration caused by the interconnects, thus high fill
factors are available.

     Probably the most interesting devices are the CMT hybrid
structures which should allow operation in both 3-5 μm and 8-14 μm
bands.   The diodes are produced by a variety of techniques, including
ion implantation and diffusion.   However, although excellent results

Fig. 5.18 Photograph of a 32 x 32 array of CMT, 8-14 μm
        photodiodes hybridized onto a silicon chip (Courtesy
        of Mullard Ltd).

have been achieved with the short wavelength devices, full staring
operation of 8-14 μm arrays has not yet been demonstrated. This is
due to the difficulty of achieving high direct injection into the CCD,
which requires the photodiode impedance to be greater than that of the
CCD, typically zero bias resistance-area products (RoA) of greater
than five ohm-cm^2 are needed. Also the charge storage capacity in the
silicon limits the stare time to approximately 20 μsec unless some of
the background signal can be removed.

However, Ballingall et al (1982) have produced a scheme in which
the multiplexer operation is carried out by MOS switches, instead of a
CCD. The scheme is illustrated in Figure 5.17, which shows each
detector with its anode connected to a common ground line and the
cathode connected through a switch to an output rail. The elements
are switched, a line at a time, and the output signals integrated and
multiplexed to produce a single video line. A photograph of a 32 x
32 array of CMT diodes connected by MOSFET switches to a silicon
substrate is shown in Figure 5.18. This approach has enabled two
dimensional CMT arrays to be used for imaging in the 8-14 μm band with
excellent results as shown in Figure 5.19. However, as there is no
storage for each pixel and integration only occurs for a line sample

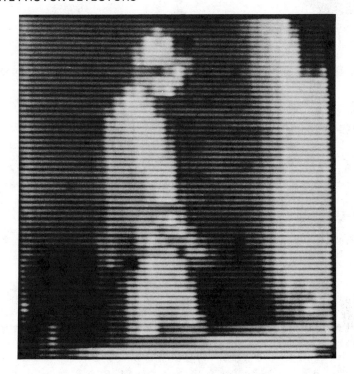

Fig. 5.19 Photograph of a thermal image obtained using an
8-14 µm, 32 x 32 line scanned array.

time the performance of this array is only equivalent to that which
would be obtained with a scanned linear array, but the requirement of
a mechanical scanner has been eliminated.

The use of these arrays for staring thermal imaging systems is
unlikely to be their most important application initially, which will
probably be in missile guidance systems requiring large numbers of
arrays at relatively low costs.

5.7 Applications of Photon Detectors

Photon detectors have been used in a wide range of applications
including the military, industrial and medical markets. Many of the
systems are common to all fields and the requirements have existed for
many years, the recent developments being due to new detectors
offering improved sensitivity. The most obvious application of an
infrared detector is as a heat seeker, which might produce an image of
the object or range and radiometric information.

The military market is by far the largest, and vast sums have been
invested by many nations aimed at 'seeing in the dark'. However, the
benefit of this work is now being realised in medical and industrial

applications, where the money is not always available to support the fundamental research required. Simple thermal imaging systems were employed during World War II, and due to the fundamental benefit of being a passive system, work is still continuing on the development of more sophisticated systems, with temperature resolutions of a fraction of a degree. As described in Section 5.5.1 the CMT SPRITE detector has recently produced some of the best quality imaging ever seen, and systems are now being fitted to many vehicles to give them a twenty four hour fighting capability. Imaging systems are now often used for diagnostic aids in hospitals as small temperature differences, due to higher metabolic rates are observed in cases of breast cancer and rheumatoid arthritis, and for the determination of the extent of burns and frostbite. Surveillance and security systems are now marketed commercially although for most requirements the high performance obtained with photon detectors is not essential and pyroelectric devices offer a significant cost saving.

A more recent use of thermal imaging has been in conjunction with satellites for surveys of the earths resources and weather forecasting. Information is obtained on pollution of the seas, forest fire detection and geological mapping, the weather pattern and movement of storm centres can be accurately tracked and suitable precautions taken of any imminent adverse weather conditions. A development from the imaging system is the missile guidance seekers which are now available, these allow an aircraft to engage in a 'fire-and forget' encounter, as after launching the missiles they then home in on the hot spots which were designated before launch. This application is now financing much of the development of the staring arrays described in section 5.6, as the elimination of a mechanical scanner is of great benefit to these systems.

Photon detectors are not always used in passive systems, they often operate in conjunction with a laser, which enables information on range and speed to be obtained from the detection of the laser return pulse. In addition some missile systems operate with a beam rider in which a laser is aligned with a detector mounted on the missile and the missile steered by a trailing wire or radio link.

The use of photon detectors in many diverse projects is now being accepted and with further improvement and reduction in cost this will increase substantially. One problem which must be overcome to allow their use in many applications is the elimination of having to cool the detector, often to 77K for 8-14 μm operation, and there is much work aimed at this and over the next few years it is quite probable that good CMT 8-14 μm band arrays operating at room temperature will become available.

CHAPTER 6

CONCLUSIONS

During the last two decades there have been many significant
developments with photodetectors, which have enabled high
performance devices to exist throughout the visible and infrared
regions of the spectrum, offering the possibilities of extreme
sensitivities or lower cost medium performance sensors. With the
development of the NEA photocathodes, photomultiplier tubes with much
greater sensitivities and extended wavelength operation into the near
infrared are readily available. Pyroelectric devices offer broad
band detection and operate at room temperature. They are relatively
cheap and have generally replaced other thermal devices for routine
laboratory uses and are marketed in large quantities for surveillance
systems. The major developments with solid state photon detectors
have resulted in increased operating temperatures with the
introduction of intrinsic devices for long wavelength applications
which have replaced the extrinsic photoconductors, and the development
of large two dimensional staring arrays.

However, due to the large diversity of photodetectors available it
is very difficult to make a direct comparison as the application,
whether it be for imaging or radiometory, and the environment must be
a major factor in determining the correct device. If a broad band
sensor of average sensitivity is required a thermal detector is the
most ideal, as it has a flat spectral response over a large
wavelength range and can generally be operated at room temperature.
For nuclear particle and very low light intensity detection and
imaging in the visible and near infrared regions, a photomultipler
tube with the correct photocathode will be most suitable. For very
high performance applications and if cooling is acceptable the modern
solid state photon detectors offer a large selection of devices
depending on the exact wavelength of operation.

   A summary of the performance parameters of the various detectors
described in this book are shown in Table 6.1 However, it must be
remembered that a spread in performance often occurs between various
samples and the values shown are generally for high quality devices,
but a comparison of the figures will give some indication of the
relative performance.

Table 6   Detector Performances

Detector	Mode of Operation	Cut-off Wavelength (μm)	Peak Wavelength (μm)	Operating Temp (K)	Responsivity	D* (cm Hz$^{\frac{1}{2}}$ W^{-1})	Response Time (μS)
Solid State Photon Devices							
Si	PV	1.1	0.8	295		2×10^{12}	1000
GaAs	PV	0.9	0.8	295		4.5×10^{11}	1000
Tl$_2$S	PC	1.1	0.9	295		2×10^{12}	530
Ga	PV	1.8	1.6	295		4×10^{11}	
PbS	PC	2.5	2.3	295	1 AW^{-1}	1×10^{11}	300
		3.5	2.8	193	5 AW^{-1}	6×10^{11}	1000
		4.0	3.0	77	1 AW^{-1}	1×10^{11}	1000
PbSe	PC	5	3.8	295	3×10^3 VW^{-1}	1×10^{10}	2
		6	4.8	193	6×10^3 VW^{-1}	2×10^{10}	30
		7.5	5.0	77	9×10^3 VW^{-1}	3×10^{10}	40
PbTe	PC	5.1	4.0	77		3×10^9	25

... Continued

Solid State Photon Devices

$Pb_{1-x}Sn_xTe$	PV	13	11.0	77	$5 \times 10^2 \ VW^{-1}$	5×10^{10}	0.1
$Pb_{1-x}Sn_xSe$	PV	13	11.5	77		3×10^9	0.2
$InSb$	PC	7.3	6.5	295		4×10^8	0.2
		6.5	5.5	195	$1 \times 10^5 \ VW^{-1}$	1×10^{10}	0.5
		5.6	5.3	77		6×10^{10}	
$InSb$	PV	5.6	5.3	77		7×10^{10}	1
$InAs$	PC	3.8	3.6	295		1×10^8	0.2
		3.5	3.2	193		3×10^{11}	0.5
$InAs$	PV	3.7	3.4	295		7×10^9	2
		3.5	3.2	193		7×10^{10}	1
		3.4	3.0	77		7×10^{11}	1
$Hg_{1-x}Cd_xTe$	PC	5.5	4.9	295		4×10^9	0.3
		5.5	5	193		4×10^{10}	0.2
		13	11	77	$1 \times 10^4 \ VW^{-1}$	5×10^{10}	0.1
		27	20	77		1×10^{10}	
		36	30	20		6×10^{10}	

··· Continued

Solid State Photon Devices						
$Hg_{1-x}Cd_xTe$	PV	12	10	77		1×10^{10}	0.005
Ge-Au	E-PC	11	5	77	10^4 VW^{-1}	2×10^{10}	0.1
Ge-Hg	E-PC	14	10	30	10^4 VW^{-1}	3×10^{10}	0.1
Ge-Cu	E-PC	27	20	20		3×10^{10}	0.1
Ge-Zn	E-PC	40	36	4		2×10^{10}	0.01
Ge-Ga	E-PC	50	100	4	10^4 VW^{-1}	3×10^{11}	0.5
Thermal Detector						
Thermopile-semiconductor	50		300	5 VW^{-1}	3×10^9	10000
Thermopile-metal film on heat sink	50		300	5×10^{-6} VW^{-1}	1×10^6	0.03
Thermopile-thin film on polymer backing	50		300	10^2 VW^{-1}	3×10^8	100
Bolometer-thermistor	300		300	10^3 VW^{-1}	2×10^8	1000

... Continued

Detector					
Thermal Detector					
Bolometer-cooled Ge	1000	4	2×10^4 VW^{-1}	7×10^{-13} (NEP)	300
Golay Cell	1000	300	10^6 VW^{-1}	2×10^{-10} (NEP)	10000
Pyroelectric-TGS	300	300	10^3 VW^{-1}	1×10^9	1000
Pyroelectric-LiTaO$_3$	500	300	10^6 VW^{-1}	1×10^8	0.1 (with suitable electronics)
Photocathodes					
Ag-0-Cs (S1)	1.2	300	60 μA/1m		
Cs$_3$Sb (S11)	0.6	300	80 μA/1m		
Na$_2$KSb(Cs)(S20)	0.9	300	500 μA/1m		
GaAs(CsO)(NEA)	0.9	300	2000 μA/1m		
InAs$_{1-x}$P$_x$	1.1	300	1100 A/1m		

NOTES: PV – photovoltaic
         PC – photoconductive
      E-PC – extrinsic photoconductive

REFERENCES

* Selected references abstracted

Adams, J., Manley, B.W., IEEE Trans. Nucl. Sci., NS-13, 3, 88, (1966).

* Ameurlaine, J., Motte, C., Riant, Y., Royer, M., Int. Electron Devices Mtg. (Washington), p85, (1973).

* Antypas, G.A., Edgecumbe, J., Appl. Phys. Lett., 26, 371, (1975).

* Antypas, G.A., Moon, R.L., James, L.W., Edgecumbe, J., Bell, R.L., Proc. 4th Int. Conf. on GaAs and related compounds. p48, (1972), (Inst. of Physics).

* Arens, J.F., Lamb, G.M., Peck, M.C., Opt. Eng. 22, 267, (1983).

Astheimer, R.W., Schwarz, F., Appl. Opt. 7, 1687, (1968).

Astheimer, R.W., Wormser, E,M., J. Opt. Soc. Am, 49, 184, (1959).

* Ballingall, R.A., Blenkinsop, I.D., Elliott, C.T., Baker, I.M., Jenner, D., Elect. Letts. 18, 285, (1982).

* Barbe, D.F., Proc. IEEE., 63, 38, (1975).

Bartlett, B.E., Deans, J., Ellen, P.C., J. Matls. Sci., 4, 266, (1969).

* Becla, P., Pawlikowski, J.M., I.R. Phys., 16, 457, (1976).

* Bell, R.L., Negative Electron Affinity Devices, (1973), (Oxford Univ. Press).

* Bellavance, D.W., Johnson, M.R., J. Elect. Matls., 5, 363, (1976).

Birks, J.B., The Theory and Practice of Scintillation Counting, (1964), (Pergamon Press, New York).

* Bishop, S.G., Moore, W.J., Appl. Opt., $\underline{12}$, 80, (1973).

Blattner, D.J., Johnson, H.C., Ruedy, J., Sterzer, F., RCA Rev., $\underline{26}$, 22, (1965).

Bode, D.E., Physics of Thin Films, $\underline{3}$, 275, (1966), (Academic).

* Borrello, S., Kinch, M., LaMont, D.I., I.R. Phys., $\underline{17}$, 121, (1977).

Boyle, W.S., Rodgers, K.F., J. Opt. Am., $\underline{49}$, 66, (1959).

* Bryan, E., I.R. Phys., $\underline{23}$, 341, (1983).

* Bubalac, L.O., Tennant, W.E., Shin, S.H., Wang, C.C., Lanir, M., Gertner, E.R., Marshall, E.D., Japn. J. Appl. Phys., $\underline{19}$, Suppl. 19-1, 495, (19⌒⌒).

* Buckingham, M.J., Faulkner, E.A., Radio and Electronic Engineer., $\underline{44}$, 125, (1974).

Bullock, T.H., Barrett, R., Comm. in Behav. Biol. $\underline{A1}$, 19, (1968).

Cady, W.G., Piezoelectricity, (1946), (McGraw Hill, New York).

Cairns, R.B., Samson, J.A.R., J. Opt. Soc. of America, $\underline{56}$, 1568, (1966).

Calawa, A.R., Harman, T.C., Finn, M., Youtz, P., Trans. Metall. Soc. AIME., $\underline{242}$, 374, (1968).

Campbell, N.R., Phil. Mag., $\underline{12}$, 173, (1931).

* Capasso, F., J. Vac. Sci. Technl., $\underline{B2}$, 457, (1983).

Carr, W.N., Miller, E.A., Leezer, J.F., Rev. Sci. Inst., $\underline{37}$, 83, (1966).

Cashman, R.J., Proc. Inst. Radio Engrs., $\underline{47}$, 1471, (1959).

* Ceckowski, D.H., Eberhardt, E., Carney, E., IEEE Trans. on Nucl. Sci., $\underline{NS-28}$, 677, (1981).

* Chatanier, M., Gauffre, G., IEEE Trans. Inst. and Meas., $\underline{IM22}$, 179, (1973).

* Chia, P.S., Balon, J.R., Lockwood, A.H., Randall, D.M., Renda, F.J., DeVaux, L.H., Kimura, H., I.R. Phys., $\underline{15}$, 279, (1975).

* Chin, M.A., J. Appl. Phys., $\underline{48}$, 2723, (1977).

* Clarke, J., Richards, P.L., Yeh, N.H., Appl. Phys. Lett., 30, 664, (1977).

* Cohen-Solal, G., Sella, C., Imhoff, D., Zozime, A., Japan. J. Appl. Phys., Suppl. 2 Pt.1, 517, (1974).

* Cohen-Solal, G., Zozime, A., Motte, C., Riant, Y., I.R. Phys., 16, 555, (1976).

* Contreras, B., Gaddy, O.L., Appl. Phys. Letts., 18 277, (1971).

* Coron, N., I.R. Phys., 16, 411, (1976).

* Cuthbertson, G.M., MacGregor, M.P., Proc. Int. Conf. on Advanced Detectors and Systems., London 1981, IEE Conf. Publ., 204, 30, (1981).

D'Asaro, L.A., Anderson, L.K., Electronics 39, 94, (1966).

Dennis, P.N.J., Private Communication, (1978).

* Dhawan, S., IEEE Trans. on Nuclear Sci., NS-28, 672, (1981).

* Draine, B.T., Sievers, A.J., Opt. Comm., 16, 425, (1976).

Drew, H.D., Sievers, A.J., Appl. Opt., 8, 2067, (1969).

* Drummond, A.J., Advan. In Geophysics., 14, 1, (1970).

Electronic Industries Association., Joint Electron Devices Engineering Council., Publication No 50, (1964).

* Elliott, C.T., Handbook of Semiconductors 4, 727, (1981).

* Elliott, C.T., Electron. Letters, 17, 312, (1982).

* Elliott, C.T., Day, D., Wilson, D.J., I.R. Phys., 22, 31, (1982).

Elster, J., Geitel, H., Ann. Physik., 38, 497, (1889).

* Emmons, R.B., I.R. Phys., 10, 63, (1970).

* Escher, J.S., Antypas, G.A., Edgecumbe, J., Appl. Phys. Lett., 29, 153, (1976).

* Ewing, W.S., Shepherd, F.D., Capps, R.W., Dereniak, E.L., Optical Eng. 22, 334, (1983).

Fellgett, P.B., Radio and Electron Eng., 42, 476, (1972).

* Fiorito, G., Gasparrini, G., Svelto, F., Appl. Phys. Letts., 23, 443, (1973).

* Fiorito, G., Gasparrini, G., Svelto, F., I.R. Phys., 15, 287, (1975).

* Fiorito, G., Gasparrini, G., Svelto, F., Appl. Phys. 17, 105, (1978).

* Fisher, D.G., Enstron, R.E., Escher, J.S., Williams, B.F., J. Appl. Phys., 43, 3815, (1972).

Fizeau, H.L., Foucault, L., Comptes Rendus, 25, 447, (1847).

* Foyt, A.G., Harman, T.C., Donnelly, J.P., Appl. Phys. Letts., 18, 321, (1971).

* Foyt, A.G., Lindley W.T., Donnelly, J.P., Appl. Phys. Letts., 16, 335, (1970).

Galavanov, V.V., Zlatkin, L.V., Inst. and Experimental Techniques, No 6, 1471, (1965).

* Gallinaro, G., Varone, R., Cryogenics, 15, 292, (1975).

Gebbie, H.A., Harding, W.R., Hilsum, C., Pryce, A.W., Roberts, V., Proc. Roy. Soc., A206, 87, (1951).

* Ghosh, C., Proc. SPIE., 346, 62, (1982).

Golay, M.J.E., Rev. Sci. Inst., 18, 357, (1947).

Goodrich, G.W., Riley, W.C., Rev. Sci. Inst., 33, 761, (1962).

Gorlich, P., Z. Physik, 101, 335, (1936).

* Groves, S.H., J. Electronic Materials, 6, 195, (1977).

Hadni, A., J. Phys., 24, 694, (1963).

* Hadni, A., Thomas, R., Mangin, J., Bagard, M., I.R. Phys., 18, 663, (1978).

Hallwachs, W., Ann. Physik., 33, 301, (1888).

* Hamilton, C.A., Phelan Jr, R.J., Day, G.W., Opt. Spectra, 9, 37, (1975).

* Harman, T.C., J. Non Metals, 1, 183, (1973).

* Hauser, M.G., Notarys, H.A., Bull. Am. Astro. Soc., 7, 409, (1975).

Heroux, L., Hinteregger, H.E., Rev. Sci, Inst., 31, 280, (1960).

Herschel, W., Trans. Roy. Soc., (London), 90, Pt II, 255, (1800).

Hertz, H., Ann. Physik, 31, 983, (1887).

Hickey, J.R., Daniels, D.B., Rev. Sci. Inst., 40, 732, (1969).

* Hodges, D.T., McColl, M., Appl. Phys. Lett. 30, 5, (1977).

* Hohnke, D.K., Holloway, H., Appl. Phys. Letters, 24, 633, (1974).

Hudson, R.D., I.R. System Engineering, (1969), (Wiley and Sons, New York).

* Igras, E., Piotrowski, J., Zimmoch-Higersberger, I., Electron Technology, 10, 63, (1977).

* James (1), L.W., Antypas, G.A., Edgecumbe, J., Moon, R.L., Bell, R.L., J. Appl. Phys., 42, 4976, (1971).

* James (2), L.W., Antypas, G.A., Uebbing, J.J., Yep. T.O., Bell, R.L., J. Appl. Phys., 42, 580, (1971).

* Johnson, E.S., Schmit, J.L., J. of Electronic Materials, 6, 25, (1977).

Jones, R.C., Advances in Electronics, 5, 1, (1953), (New York, Academic).

Kimmitt, M.F., Prior, A.C., J. Electrochem. Soc., 108, 1034, (1961).

* Kinch, M.A., J. Appl. Phys., 42, 5861, (1971).

* Kinch, M.A., Borrello, S.R., Breazale, B.H., Simmons, A., I.R. Phys. 17, 137, (1977).

* Kohn, E.S., Roosild, S.A., Roosild, S.A., Shepherd, F.D., Yang, A.C., Proc. Int. Conf. on Applic. of CCDs, (San Diego), p59, (1975).

Kohn, A.N., Schlickman, J.L., IEEE., Trans. Electron. Devices, ED-16, 885, (1969).

Koller, L.R., J. Opt. Soc. Amer., 19, 135, (1929).

Koller, L.R., Phys. Rev., 36, 1639, (1930).

* Kosonocky, W.F., Elabd, H., Erhardt, H.G., Shallcross, F.V., Meray, G.M., Villani, T.S., Groppe, J.V., Miller, R., Frantz, V.L., Cantella, M.J., Klein, J., Roberts, N., Proc. SPIE. 344, 66, (1982).

* Krall, H.R., Helvy, F.A., Persyk, D.E., IEEE Trans. Nucl. Sci., NS-17, 71, (1970).

Kruse, P.W., J. Appl. Phys., 30, 770, (1959).

* Kruse, P.W., Topics in App. Phys., 19, 5, (1977). (Ed Keyes, R.J., Springer Verlag, New York).

Kruse, P.W., McGlauchlin, L.D., McQuistan, R.B., Elements of I.R. Technology, (1962) (Wiley and Sons, New York).

Kurnick, S.W., Zitter, R.N., J. Appl. Phys., 27, 278, (1956).

* Kutscher, E.W., Laser 75 - Optoelectronics Conference Proceedings, p206, (1975).

Langley, S.P., Nature, 25, 14, (1881).

Langley, S.P., Proc. American Acad. Arts and Sci., 16, 342, (1881).

Lawson, W.D., Nielsen, S., Putley, E.H., Young, A.S., J. Phys. Chem. Solids, 9, 325, (1959).

* Lock, P.J., Appl. Phys. Letts., 19, 390, (1971).

Loh, E., J. Phys. and Chem. Solids, 24, 493, (1963).

Long, D., I.R. Phys., 7, 121, (1967).

Long, D., I.R. Phys., 7, 169, (1967).

* Long, D., Schmit, J.L., Semiconductors & Semimetals, 5, Chap 5, (1970), (Ed Willardson and Beer, Academic Press New York).

* Longo, J.T., Cheung, D.T., Andrews, A.M., Wang, C.C., Tracy, J.M., IEEE Trans. on Electron Devices, ED-25, 213, (1978).

Low, F.J., J. Opt. Soc. Am., 51, 1300, (1961).

Low, F.J., Proc. IEEE, 53, 516, (1965).

Low, F.J., Proc. IEEE, 54, 477, (1966).

Lucovsky, G., Emmons, R.B., Proc. IEEE, 53, 180, (1965).

* Marine, J., Motte, C., Appl. Phys. Letts., 23, 450, (1973).

Martin, D.H., Bloor, D., Cryogenics, $\underline{1}$, 159, (1961).

Martinelli, R.U., Appl. Phys. Lett., $\underline{16}$, 261, (1970).

* Martinelli, R.U., Fisher, D.G., Proc. IEEE $\underline{62}$, 1339, (1974).

* McColl, M., Hodges, D.T., Garber, W.A., IEEE Trans. Microware Theory Tech., $\underline{MTT-25}$, 463, (1977).

* McNally, P.J., Radiation Effects, $\underline{6}$, 149, (1970).

Melloni, M., Ann. Phys. $\underline{28}$, 371, (1833).

* Melngailis, I., Harman, T.C., Semiconductors and Semimetals, $\underline{5}$, Chap 4, (1970). (Eds. Willardson and Beer, Academic Press New York).

Morten, F.D., King, R.E.J., Appl. Optics, $\underline{4}$, 659, (1965).

Morton G.A., Appl. Optics, $\underline{7}$, 1, (1968).

* Moss, T.S., Burrell, G.J., Ellis, B., Semiconductor Opto-Electronics, (1973), (Butterworths London).

* Moustakas, T.D., Connell, G.A.N., J.A.P., $\underline{47}$, 1322, (1976).

Nakamura, J.K., Schwarz, S.E., Appl. Opt., $\underline{7}$, 1073, (1968).

* Nayar, P.S., I.R. Phys., $\underline{14}$, 31, (1974).

Nobili, L., Ann. Physik, Ser. 2. $\underline{20}$, 245, (1830).

* Nudelman, S., Electronic Imaging Conf., London 11-13 Sept., (1978), p253, (Academic Press)

* Obak, K., Rehak, P., IEEE Trans. on Nucl. Sci. $\underline{NS-28}$, 683, (1981).

* Olsen, G.H., Szostak, D.J., Zamerowski, T.J., Ettenberg, M., J. Appl. Phys., $\underline{48}$, 1007, (1977).

* Polla, D.L., Sood, A.K., J. Appl. Phys. $\underline{51}$, 4908, (1980).

* Porter, S.G., Ferroelectrics, $\underline{33}$, 193, (1981).

Putley, E.H., Proc. Phys. Soc., $\underline{76}$, 802, (1960).

Putley, E.H., J. Phys. Chem. Solids, $\underline{22}$, 241, (1961).

Putley (1), E.H., I.R. Phys. $\underline{4}$, 1, (1964).

Putley (2), E.H., Phys. Stat. Solidi., 6, 571, (1964).

Putley, E.H., J. Sci. Instr., 43, 857, (1966).

Putley, E.H., Optical Properties of Solids, p175, (1969) (Plenum Press New York).

* Putley, E.H., Semiconductors and Semimetals, 5, Chap 6, (1970), (Eds Willardson and Beer Academic Press New York)

* Putley, E.H., Semiconductors and Semimetals, 12, Chap 3, (1977), (Eds Willardson and Beer, Academic Press, New York).

* Putley, E.H., Topics in Appl. Phys., 11, 71, (1977).

* Putley, E.H., Ferroelectrics, 33, 207, (1981).

* Reine, M.B., Sood, A.K., Tredwell, T.J., Semiconductors and Semimetals, 18, Chap 6, (1981), (Eds. Willardson and Beer, Academic Press New York).

* Richards, P.L., Clarke, J., Hoffer, G.I., Nishioka, N.S., Woody, D.F., Yeh, N.H., 2nd Int. Conf. on Sub. mm. Waves & their Applic., (1976) p64, (Puerto Rico) (IEEE N Y).

Ridgway, S.L., Design Electronics, 5, 20, (1968).

Ritter, J.W., Ann. Physik, 12, 409, (1803).

* Roberts, C.G., Proc. SPIE, 443, 131, (1983).

* Rolls, W.H., Eddolls, D.V., I.R. Phys., 13, 143, (1973).

* Roundy, C.B., Byer, R.L., Phillson, D.W., Kuizenga, D.J., Opt. Comm., 10, 374, (1974).

Ruggles, P.C., Slark, N.A., IEEE. Trans. on Nucl. Sci., NS-11, 100, (1964).

Ryvkin, S.M., Photoelectric Effects in Semiconductors, Consultants Bureau, New York, (1964).

Scheer, J.J., Van Laar, J., Solid State Common., 3, 189, (1965).

Schmit, J.L., Steltzer, E.L., J. Appl. Phys., 40, 4865, (1969).

Schwarz, E., Research, 5, 407, (1952).

* Sclar, N., I.R. Phys., 16, 435, (1976).

* Sclar, N., I.R. Phys., <u>17</u>, 71, (1977).

Scott, W., Kloch, A.E., U.S. Patent 3, 743, 553, (1973).

Scott-Barr, E., I.R. Phys., <u>2</u>, 67, (1962).

* Shin, S.H., Vanderwyck, A.H.B., Kim, J.C., Cheung, D.T., App. Phys.
Letts., <u>37</u>, 402, (1980).

* Shivanandan, K., McNutt, D.P., Bell, R.J., I.R. Phys., <u>15</u>, 27,
(1975).

Simon, R.E., Sommer, A.H., Tietjen, J.J., Williams, B.F., Appl.
Phys. Lett., <u>13</u>, 355, (1968).

Simon, R.E., Spicer, W.E., J. Appl. Phys., <u>31</u>, 1505, (1960) Phys
Rev., <u>119</u>, 621, (1960).

Slawek, J.E., Proc. Tech. Colloq. on Cryogenics & I.R. Detection,
p171, (1969) (Eds Hogan & Moss, Boston Tech.)

Smith, R.A., Jones, F.E., Chasmar, R.P., The Detection & Measurement
of Infra Red Radiation, (1968), (Oxford Univ. Press, London).

Sommer, A.H., Photoemissive Materials: Preparation Properties and
Uses, (1968) (Wiley, New York).

* Sommer, A.H., J. Vac. Sci. Technol. <u>Al(2)</u>, 119, (1983).

Soreff, R.A., J. Appl. Phys., <u>38</u>, 5201, (1967).

Spicer, W.E., Phys. Rev., <u>112</u>, 114, (1958).

Spicer, W.E., J. Appl. Phys., <u>31</u>, 2077, (1960).

* Spicer, W.E., Appl. Phys., <u>12</u>, 115, (1977).

* Stevens, N.B., Semiconductors and Semimetals, <u>5</u>, Chap 7, (1970)
(Eds Willardson and Beer, Academic Press, New York).

* Stillman, G.E., Wolfe, C.M., Dimmock, J.O., Proc. Symp. Sub
millimetre Waves, p345, (1971) (Polytechnic Inst. of Brooklyn, New
York).

* Stillman, G.E., Wolfe, C.M., Dimmock, J.O., Semiconductors and
Semimetals, <u>12</u>, Chap 4, (1977), (Eds Willardson and Beer, Academic
Press, New York).

* Stokowski, S.E., Venables, J.D., Byer, N.G., Ensign, T.C., I.R.,
Phys., <u>16</u>, 331, (1976).

* Stotlar, S.C., McLellan, E.J., Gibbs, A.J., Webb, J., Ferroelectrics, 28, 325, (1980).

Strauss, A.J., Trans. Metall. Soc. AIME 242, 354, (1968).

* Stupp, E.H., SPIE, Technical Symposium East, Reston Virginia, 22-25 March 1976, Proc. SPIE 78, 23, (1976).

* Thomas, C.M., Proc. SPIE 42, 71, (1973).

Trautwein, J.W., Electronic Industries, 23, 72, (1964).

Van der Ziel, A., Proc. IRE, 43, 1639, (1955).

Van der Ziel, A., Becking, A.G.T., Proc. IRE, 46, 589, (1958).

Van der Ziel, A., Noise in Electron Devices, (1959) (L.D. Smullin & H.A. Haus, John Wiley & Sons, New York).

Van Vliet, K.M., Appl. Opt., 6, 1145, (1967).

Verie, C., Ayas, J., Appl. Phys. Letts., 10, 241, (1967).

Verie, C., Granger, R., C.R. Acad. Sci., Paris, 261, 3349, (1965).

* Von Ortenberg, M., Link, J., Helbig, R., J. Opt. Soc, Am, 67, 968, (1977).

* Wang, C.C., Lorenzo, J.S., I.R. Phys., 17, 83, (1977).

* Warner, D.J., Pedder, D.I., Moody, I.S., Burrage, J., Ferroelectrics, 33, 249, (1981).

* Watton, R., Ferroelectrics, 10, 91, (1976).

* Watton, R., Manning, P., Burgess, D., I.R. Phys., 22, 259, (1982).

* Watton, R., Manning, P., Burgess, D., Proc. SPIE, 395, 78, (1983).

Weissler, G.L., Jap. J. Appl. Phys., 4, Suppl. 1, 486, (1965).

Williams, B.F., Simon, R.E., Phys. Rev. Lett., 18, 485, (1967).

* Williams, G.F., Capasso, F., Tsang, W.T., IEEE Electron Device Letters, EDL-3, 71, (1982).

* Woodhead, A.W., Vacuum, 30, 539, (1980).

Wooten, F., Spicer, W.E., Surf. Sci, 1, 367, (1964).

Yamaka, E., Teranishi, A., Nakamura, K., Nagashima, T., Ferro-electrics, <u>11</u>, 305, (1976).

Yeou, Ta, Compt-Rend, <u>207</u>, 1042, (1938).

* Yokozawa, F., JEE (Japan), <u>19</u>, 94, (1982).

Zwerdling, S.,. Smith, R.A., Thericult, J.P., I.R. Phys., <u>8</u>, 271, (1968).

* Zwicker, H.R., Optical and I.R. detectors. Ch 3, 149, (1977) (R.J. Keyes, Springer-Verlag New York).

SELECTED ABSTRACTS

Ameurlaine, J., Motte, C., Riant, Y., and Royer, M.

RECENT PROGRESS IN THE FABRICATION OF PHOTOVOLTAIC $Hg_{1-x}Cd_xTe$ DETECTORS

The $Hg_{1-x}Cd_xTe$ ternary compound is a material in which the forbidden bandgap can be varied. It lends itself ideally to the fabrication of photovoltaic infrared detectors in the 2 to 12 μm spectral region. Special effort was directed toward the development of a component with superior performance and reliability. The crystal growing process is controlled to yield large diameter, homogeneous crystals. Mass production techniques commonly used in the semiconductor industry are implemented for the processing of the photodiodes. More advanced techniques, such as ion implantation are being considerd. Single element detectors and multi-element arrays with high performance characteristics in D*, quantum efficiency and cut-off frequency have been fabricated.

Antypas, G. A., and Edgecumbe, J.

GLASS-SEALED GaAs-AlGaAs TRANSMISSION PHOTOCATHODE

A GaAs/GaAlAs/GaAs/GaAlAs heterostructure has been prepared on a GaAs substrate, bonded to 7056 Corning glass, and the substrate and first AlGaAs removed chemically, utilizing the differential etching characteristics of GaAs and AlGaAs in $NH_4OH-H_2O_2$ and HF solutions.

127

The resulting structure of GaAs/AlGaAs/glass has excellent layer
morphology, uniform thickness, and good transmission photocathode
performance.

Reprinted with permission from Applied Physics Letters, Vol. 26,
No. 7, pp 371, (1975).

Antypas, G. A., Moon, R L., James, L. W., Edgecumbe, J. and Bell, R.

III-V QUATERNARY ALLOYS (HETEROJUNCTIONS AND PHOTOEMISSIVE DEVICES)

The unqiue relationship between lattice constant and bandgap in
ternary semiconductors limits the application of these compounds, for
heterojunction devices, to lattice-matched systems, specifically to
(Ga-Al) group V alloys. The extension of heterojunction devices to
other III-V compounds, however, can be accomplished by the addition of
a fourth component, introducing an extra degree of freedom, thus
permitting the independent variation of lattice constant and bandgap.
Two quaternary systems were investigated. GaAlAsSb and InGaAsP.
Liquid phase epitaxial layers were grown on GaAs and InP respectively.
Growth parameters were determined for both systems, to yield lattice
matched heterostructures over a wide energy range. Measurement on p-n
junctions and photoemissive devices indicate material quality similar
to that of the respective ternaries.

Proc. of the 4th Int. Symposium on Gallium Arsenide and related
Compounds, Boulder, USA, pp 48, (1972).

Arens, J. F., Lamb, G. M., and Peck, M. C.,

INFRARED CAMERA FOR TEN MICROMETER ASTRONOMY

An infrared imaging photometer employing a monolithic 32 x 32
pixel bismuth doped silicon charge injection device array is
described. The device is primarily useful in the 8 to 13 μm
atmospheric window. The detector is sufficiently sensitive to provide
good performance on ground-based telescopes and promises to be very
good for low background space flight operation.

Reprinted with permission from Optical Eng., Vol. 22 No. 2, pp 267,
(1983).

Ballingall, R. A., Blenkinsop, I. D. and Elliott, C. T.,

ELECTRONICALLY SCANNED CMT DETECTOR ARRAY FOR THE 8-14 μm

A cadmium-mercury-telluride photovoltaic hybrid array of 1024

elements is described. The device operates in the 8-14 μm band and is
electronically scanned by means of MOS silicon switches. An example
of a 'staring' image obtained with the device is shown.

Barbe, D. F.,

IMAGING DEVICES USING THE CHARGE-COUPLED CONCEPT

A unified treatment of the basic electrostatic and dynamic design
of charge-coupled devices (CCD's) based on approximate analytical
analysis is presented. Clocking methods and tradeoffs are discussed.
Driver power dissipation and on-chip power dissipation are analyzed.
Properties of noise sources due to charge input and transfer are
summarized. Low-noise methods of signal extraction are discussed in
detail. The state of the art for linear and area arrays is presented.
Tradeoffs in area-array performance from a systems point of view and
performance predictions are presented in detail. Time delay and
integration (TDI) and the charge-injection device (CID) are discussed.
Finally, the uses of the charge-coupled concept in infrared imaging
are discussed.

Becla, P., and Pawlikowski, J. M.,

EPITAXIAL $Cd_xHg_{1-x}Te$ PHOTOVOLTAIC DETECTORS

Photovoltaic p-n junction detectors made of epitaxial $Cd_xHg_{1-x}Te$
layers have been constructed with detectivities (at $77^\circ K$) greater than
10^{11} cm Hz$^{\frac{1}{2}}$ W^{-1} at 2.6 μm and up to 10^{10} cm Hz$^{\frac{1}{2}}$ W^{-1} at 9-12 μm.
Junctions with a peak of photoresponse at 2-8 μm have detectivities
near the 180° FOV background limited value. Measurements from 10Hz to
1.3 kHz have shown that 1/f noise prevails for f $\leqslant$ 100 Hz whereas the
Johnson noise is predominant for higher frequencies. The detectors
obtained have quantum efficiencies up to 70% at peak of
photosensitivity and response times 10-300 nsec depending on the
wavelength of the incident radiation.

Bell, R. L.,

NEGATIVE ELECTRON AFFINITY DEVICES

This book provides a brief account of the new and rapidly developing technology of negative electron affinity devices. The electronic properties and diffusion of carriers in solids, and the special technology required to produce negative affinity photoemitters are reviewed. One chapter deals with materials technology including liquid and vapour phase epitaxy, molecular beam epitaxy of 3-5 compounds and results of photoemissive yields and spectral responses of many cathodes are given.

Oxford University Press, Editors P. Hammond and D. Walsh, (1973).

Bellavance, D. W., and Johnson, M. R.

OPEN TUBE VAPOR TRANSPORT GROWTH OF $Pb_{1-x}Sn_xTe$ EPITAXIAL FILMS FOR INFRARED DETECTORS

Epitaxial films of $Pb_{1-x}Sn_xTe$ have been grown by open tube vapor transport on (100) $Pb_{1-x}Sn_xTe$ substrates. The as-grown films are suitable for detector array fabrication with respect to both surface smoothness and electrical properties. Charge compositions from 1% excess metal to 1% excess Te were used. Growth rates up to 3-4 μm per hour were achieved. The as-grown carrier concentrations varied from 3×10^{16} cm^{-3} to 3×10^{17} cm^{-3} depending on growth temperature and charge composition. Schottky barrier detectors with semi-transparent electrodes were fabricated on as-grown layers with no surface preparation. Good uniformity of detector parameters was obtained with arrays of 20 to 40 elements. The array size is not limited by either substrate size or epitaxial quality. Resistance-area products on the order of 1 ohm-cm^2 were obtained at 77 K for detectors with a 12 μm long wavelength cutoff. Quantum efficiencies for 8-12 μm radiation were 40-50%. Peak response and 50% cutoff occurred at 11 and 12 μm, respectively. Uniformity of cutoff wavelength across the arrays of $\pm$ 0.1 μm was obtained.

Reprinted with permission from Journal of Electronic Materials, Vol. 5, No. 3, pp 363, (1976). a publication of The Metallurgical Society of AIME, Warrendale, Pennsylvania.

Bishop, S. G. and Moore W. J.,

CHALCOGENIDE GLASS BOLOMETERS

The chalcogenide glass $Tl_2SeAs_2Te_3$ has been evaluated as a thermister bolometer material for room temperature operation. Thin film bolometers were fabricated on mica, glass and sapphire substrates by both hot-pressing and rf sputtering techniques. Best results were achieved with 10 μ thick $Tl_2SeAs_2Te_3$ elements on thin mica substrates. Using a 500 K blackbody at a 10 Hz chopping frequency, a 2.5×10^{-3} cm^2

device yielded an NEP of 2.3 x 10^{-9} W Hz$^{-\frac{1}{2}}$, and a 10^{-4}-cm^2 device achieved an NEP of 7.7 x 10^{-10} W Hz$^{-\frac{1}{2}}$. The ac performance of these devices is limited by their inherently long response times ($\tau \simeq 1$ sec).

Borrello, S., Kinch, M. and LaMont, D.,

PHOTOCONDUCTIVE HgCdTe DETECTOR PERFORMANCE WITH BACKGROUND VARIATIONS

The variations of responsivity and g-r noise with background photon flux above 10^{16} photons cm^{-2} sec^{-1} show strong majority carrier effects for 0.1 eV HgCdTe detectors. The background dependence of Auger lifetime and excess majority and minority carriers densitites are sufficient to describe observed phenomena. Excess 1/f noise seems related to the minority carrier density resulting in an observed decrease in 1/f noise corner with reduced background photon flux.

Bryan, E

OPERATING TEMPERATURE OF EXTRINSIC Si PHOTOCONDUCTIVE DETECTORS

Extrinsic Si photoconductive detectors require a lower operating temperature than do intrinsic photoconductors of the same activation energy. This paper compares D* as a function of operating temperature, for extrinsic and intrinsic detectors based on established theoretical relations and measured data. The effects of such parameters as impurity concentration, free-carrier lifetime and capture cross section (σ) are evaluated.

The activation energy dominates the alterable physical parameters of D*(T). The larger σ of extrinsic Si is shown to be directly related to extrinsic photoconductors requiring lower operating temperatures than intrinsic materials. Of the extrinsic Si temperature-limiting parameters, only σ shows any promise as a controllable parameter. This study shows that extrinsic Si will remain limited by lower operating temperatures unless its σ is decreased substantially. Materials with small σ measured as having longer lifetimes, would operate at higher temperatures.

Bubulac, L. O., Tennant, W. E., Shin, S. H., Wang, C. C., Lanir, M., Gertner, E. R. and Marshall, E. D.,

ION IMPLANTATION STUDY OF HgCdTe

Light atom species, such as Li, Mg, B, Be, Cl, F and Al implanted in bulk and epitaxial HgCdTe of compositions from 3 to 12 µm cut-off yielded n/p junctions. The nature of these junctions has not previously been understood. Implantation of light atoms has been observed to induce n-type electrically active defects which propagate deep into the material during the implantation process. The experiments, performed on an epitaxial wafer of ~5 µm cut-off wavelength, showed an electron profile of the implanted layer 3 to 4 times deeper than the mean projected range determined by SIMS measurements. The results are in good agreement with the graded junction profile determined from C-V measurement, and have junction depths of 2-3 µm determined from EBIC measurements. It has been concluded that the junction formation mechanism in HgCdTe is determined primarily by radiation induced mobile defects.

Reprinted with permission from Proceedings of the 11th Conference (1979 International) on Solid State Devices, Tokyo, 1979, Japanese Journal of Applied Physics, Vol. 19, supplement 19-1, pp 495, (1980).

Buckingham, M. J. and Faulkner, E. A.,

THE THEORY OF INHERENT NOISE IN P-N JUNCTION DIODES AND BIPOLAR TRANSISTORS

A new calculation is made of the noise (excluding excess noise) in p-n junction diodes and bipolar transistors. The theory is based on an analysis of the current waveforms caused by thermal motion of individual carriers and by individual recombination events in a one-dimensional model; unlike that of van der Ziel (1955) it does not rely on arbitrary assumptions about thermal fluctuations in carrier concentration and does not involve a transmission-line analogy. When depletion-layer recombination effects are neglected, the results are found to be identical to those obtained by van der Ziel, although one of his postulates was invalid. The theory of can der Ziel and Becking (1958) which also leads to the same results does not appear to be consistent with the new treatment.

Reprinted with permission from The Radio and Electronic Engineer, Vol. 44, No. 3., (1974).

Capasso, F.,

BAND-GAP ENGINEERING VIA GRADED GAP, SUPERLATTICE AND PERIODIC DOPING

STRUCTURES: APLICATIONS TO NOVEL PHOTODETECTORS AND OTHER DEVICES

Recent new multilayer structures and their device applications are reviewed. These new concepts allow one to radically modify the conventional energy band diagram of a pn junction and thus tailor the high field transport properties to an unpreceedented degree (band-gap engineering). This approach has been used to propose and implement a new class of avalanche photodiodes with enhanced ionization rates ratio and the solid state analog of a photomultiplier (staircase detectors). Other device applications such as repeated velocity overshoot structures are also discussed.

Ceckowski, D. H., Eberhardt, E. and Carney, E.,

PROXIMITY FOCUSED MICROCHANNEL PLATE PHOTOMULTIPLIER TUBES

Several forms of microchannel plate photomultiplier tubes (MCP PMT) for both high speed and multiple anode output applications are discussed. These tubes are from one MCP, with moderate gain, to three MCPs for photon counting gain levels. Proximity focus maintains distortion-free imaging through the tube structure. Performance data on photocathode types, gain, background, for each tube type is given.

Chatanier, M. and Gauffre, G.,

INFRARED TRANSDUCER FOR SPACE USES

The infrared transducer devised, for space applications, at the Office National d'Etudes et de Recherches Aerospatiales (ONERA) incorporates a pneumatic thermal detector combined with a capacitance bridge. Performance data of the assembly are as follows: noise equivalent power at 40 Hz, 2.10^{-10} W/Hz$^{\frac{1}{2}}$; signal bandwidth 200 Hz; sensitive area 3 mm^2; spectrum bandwidth 0.4 to 45 = (KRS 5 window). The equipment is designed to stand up under space conditions, such as vibrations, changes in temperature and vacuum.

Chia, P. S., Balon, J. R., DeVaux, L. H., Kimura, H., Lockwood, A. H., Randall, D. M. and Renda, F. J.,

PERFORMANCE OF PbSnTe DIODES AT MODERATELY REDUCED BACKGROUNDS

In recent years, the uniformity and performance of $Pb_{0.80}Sn_{0.20}Te$ arrays at T = 78°K 2 π FOV have been reported. In this paper, the performance of $Pb_{0.80}Sn_{0.20}Te$ in the 60–80°K temperature range is reported. Data are given for a PbTe PbSnTe mesa array formed by liquid phase epitaxy as well as for a planar diffused array fabricated on bulk $Pb_{0.80}Sn_{0.20}Te$ material. In this temperature range average values of $D^* > 1 \times 10^{11}$ cm $Hz^{\frac{1}{2}}$ W^{-1} (f/5 FOV) have been obtained for both types of diodes. Other device data are also presented.

Chin, M. A.,

A FREE-FLOATING SUPERCONDUCTING BOLOMETER – A SENSITIVE HEAT-PULSE DETECTOR

A free-floating (no substrate) granular aluminium superconducting bolometer has been made and tested. Sensitivity and time-constant measurements were compared with those of similar bolometers fabricated on glass and Al_2O_3 substrates. The responses were first studied with the bolometer isolated in a vacuum chamber which was immersed in a superfluid liquid-helium bath and cooled by conduction through the contact leads, and again after incremental increases of helium gas were introduced into the sample chamber. The measurements show that for low gas pressures ($\ll 10^{-4}$ cm Hg) the floating bolometer was between one and two orders of magnitude more sensitive than identical bolometers on substrates. With increasing gas pressure, the formation of a superfluid helium film of an approximate thickness of three atomic layers caused a reduction of the floating bolometer sensitivity by nearly two orders of magnitude. The effect on the bolometers on substrates was, on the other hand, only 30%. The results provide strong evidence that the Kapitza conductance into helium occurs in the first two or three atomic layers.

Clarke, J., Richards, P. L. and Yeh, N. H.

COMPOSITE SUPERCONDUCTING TRANSITION EDGE BOLOMETER

A composite far-infrared bolometer has been constructed that uses
an aluminium film at the superconducting transition temperature of
1.3 K as the temperature-sensitive element. The film is evaporated on
one edge of a 4 x 4 mm sapphire substrate, which is coated on the
reverse side with an absorbing film of bismuth. The best bolometer
has an electrical NEP of $(1.7 \pm 0.1) \times 10^{-15}$ W Hz$^{-\frac{1}{2}}$ at 2 Hz, and a
specific detectivity D* of $(1.1 \pm 0.1) \times 10^{14}$ cm W^{-1} Hz$^{\frac{1}{2}}$. This
measured electrical NEP is within a factor 2 of the fundamental
thermal noise limit.

Reprinted with permission from Applied Physics Letters, Vol 30, No.
12, pp 664, (1977).

Cohen-Solal, G., Sella, C., Imhoff, D and Zozime, A.

STRUCTURE AND PROPERTIES OF $Cd_xHg_{1-x}Te$ FILMS

Crystalline films of $Cd_xHg_{1-x}Te$ solid solutions have been
prepared by diode sputtering in a Hg plasma. The films, 1 to 5 μm
thick are very homogeneous and oriented epitaxially on single crystal,
(111) and (100) CdTe substrates, NaCl, mica and (111) silicon. The
crystallinity and the composition from x - 0.2 up to x - 0.8 have been
studied as functions of the deposition parameters. Both n and p-type
materials are obtained using a co-sputtering technique with biased
metallic auxiliary targets. Heterojunction and homojunction have
been obtained in this way. The instrinsic quality of the films (i.e.
homogeneity in composition, high level doping and good electronic
properties) allowed us to prepare sensitive photodiode detectors for
the near and far infrared spectrum (2-14 μm) as well as bidimensional
IR detector arrays.

Reprinted with permission from The Japanese Journal of Applied
Physics, Suppl. 2, Pt. 1, pp 517, (1974).

Cohen-Solal, G., Zozime, A., Motte, C., and Riant, Y.

SPUTTERED MERCURY CADMIUM TELLURIDE PHOTODIODE

Infrared (Hg Cd) Te photodiode detectors have been prepared by
triode sputtering in a mercury plasma system. Both p and n-type
material are obtained using co-sputtering technique with biased
metallic (gold or aluminium) auxiliary targets. High sensitivity
heterojunction and homojunction diodes for the 2-14 μm spectral region
were achieved in this way. At 10.6 μm, photodiodes of approximately 2
x 10^{-3} cm^2 area, operating at 77 K, have exhibited specific
detectivities D* (10.6, 900 Hz, 1 Hz) between 1.5 and 4 x 10^{10}
cm Hz$^{\frac{1}{2}}$ W^{-1}, zero-bias resistance area products between 1.2 and 3 Ω cm^2
and external quantum efficiencies between 55 and 60%. The noise level

measured at a frequency of 10^5 Hz is around 10^{-9} V $Hz^{-\frac{1}{2}}$.

Contreras, B. and Gaddy, O. L.

HETERODYNE DETECTION AT 10.6 μm WITH THIN-FILM BOLOMETERS

Heterodyne detection at 10.6 μm using fast-response-time thin-film
bolometer infrared detectors is described. With a local oscillator
power of 1 W, the minimum detectable power is found to be 4 x 10^{-11}
W/Hz. Good agreement is found between experimental results and
theory.

Coron, N.

INFRARED HELIUM COOLED BOLOMETERS IN THE PRESENCE OF BACKGROUND
RADIATION: OPTIMAL PARAMETERS AND ULTIMATE PERFORMANCES

In this paper the optimization of the principal parameters for a
helium cooled bolometer designed for a given experiment is discussed
and particularly when the optical background from the warm optics is
important. Optimal values of diameter, thermal conductivity between
the detector element and the cryostat, time constant, and helium bath
temperature are studied. Simple formula for the calculation of photon
noise are given. The ultimate performances possible in the present
state of art for the three-part bolometer are reviewed in the
different cases and compared to the theoretical limits.

Cutherbertson, G. M. and MacGregor, A. D.

HIGH PERFORMANCE THERMAL IMAGERS WITH AUTOMATED DISPLAY

The paper describes in qualitative terms two aspects of thermal
imager design, based upon SPRITE detectors. This reduces the
electronic complexity of the system and produces better performance
than that of equipments based upon discrete detectors.

The first section of the paper deals with the design of a high
performance imager. The final part discusses the processing both

analogue and digital required to improve the visibility of information in the scene.

The Proceedings of The International Conference on Advanced Detectors and Systems, IEEE Conf. Publ. Vol. 204, pp 30, (1981).

Dhawan, S.

INTRODUCTION TO MICROCHANNEL PLATE PHOTOMULTIPLIERS

A microchannel plate has about half a million channels per square centimeter, each of which functions as an independent electron multiplier. An MCP photomultiplier containing a photocathode, MCP and an anode can be constructed in a sealed envelope. The advantages of the MCP photomultiplier over a conventional photomultiplier are better time resolutions, ability to run in magnetic fields, and imaging possibilities. In principle, the three advantages can be incorporated in a single unit. With the current state of the technology, it is best to realize any two advantages in a tube.

Draine, B. T. and Sievers, A. J.

A HIGH RESPONSIVITY, LOW-NOISE GERMANIUM BOLOMETER FOR THE FAR INFRARED

A high responsivity, low noise bolometer, for the detection of far-infrared radiation has been developed, constructed from p-type compensated germanium. The paper describes the construction and design of the device and the measurements and performance obtained. When operated at 0.1K a responsivity of 2.8×10^7 VW^{-1} and an estimated limiting NEP of 3×10^{-16} W $Hz^{-\frac{1}{2}}$ have been obtained.

Optics Communications, Vol. 16, No. 3, pp 425 (1976).

Drummond, A. J.

PRECISION RADIOMETRY AND ITS SIGNIFICANCE IN ATMOSPHERIC AND SPACE PHYSICS

A review of the properties of the electromagnetic radiation, from the ultraviolet to the far-infrared region is given in this paper. An historical review of the most important developments in thermal radiometry from the first investigations of the visible spectrum in 1666 by Newton up until the establishment of high accuracy spectral radiance standards in 1968 is presented. The author also considers the sources of extra terrestrial solar fluxes and some results achieved by direct measurements at the Eppley and Jet Propulsion Laboratories of the atmospheric vertical profile.

Advances in Geophysics, Vol. 14, pp 1 (1970).

Elliott, C. T.

INFRARED DETECTORS

With the rapid developments in infrared detection in the last thirty years a wide variety of detector types operating throughout the spectrum have been produced. The majority of these devices use semiconducting materials and a survey of the many types now available is presented. The sources of noise and fundamental performance limits are discussed and the materials used are surveyed. Finally a detailed review of the limitations and operating characteristics of some of the more widely used and some of the newly developing devices is given.

Handbook of Semiconductors, edited by T. S. Moss, Vol. 4, edited by C. H. Hilsum, North Holland Publishing Company, pp 727, (1981). Copyright © Controller HMSO, London.

Elliott, C. T.

NEW DETECTOR FOR THERMAL IMAGING SYSTEMS

A new type of infrared device based on n-type cadmium mercury telluride is described which performs both the detection function and the time delay integration function in serial or serial-parallel scan thermal imaging systems. It is a simple device which considerably reduces system complexity in either the 8-14 μm or 3-5 μm wavebands.

Reprinted with permission from Electronics Letters, Vol. 17, No. 8, pp 312, (1981). Copyright © Controller HMSO, London.

Elliott, C. T., Day, D. and Wilson, D. J.

AN INTEGRATING DETECTOR FOR SERIAL SCAN THERMAL IMAGING

The operation and characteristics are described for a new type of infrared detector which is known by the acronym SPRITE (signal processing in the element). In its simplest form the device is a three lead structure in n-type cadmium-mercury-telluride which performs the same function in a thermal imaging system as an array of serial scanned elements together with the associated preamplifiers and time-delay-integration circuits.

The operating principle and basic theory of the device are described and compared with experimental data. The emphasis is on devices operating at 77K in the 8-14 μm band though some information on 3-5 μm band operation at temperatures in the region of 200K is also presented. The principal application of the devices at present is in 8-14 μm band systems where they provide equivalent performance to arrays of approximately 100 BLIP elements but with considerable simplification of the system electronics and detector fabrication.

Emmons, R. B.

THE FREQUENCY RESPONSE OF EXTRINSIC PHOTOCONDUCTORS WITH REDUCED BACKGROUND

The frequency response of a uniformly illuminated extrinsic photoconductor at reduced background levels is shown to be $F(\omega) = (1 + \omega^2\tau^2)^{-\frac{1}{2}}(1 + \omega^2\tau_p^2 R_L^2/R_D^2)^{-\frac{1}{2}}$ where τ is the recombination time of the photoconductor and τ_p is its dielectric relaxation time. R_D and R_L are respectively the resistances of the photoconductor and its load. The derivation assumes space charge neutrality which is shown to hold in these devices at reduced background levels. When the detector and load resistances are matched, this frequency response implies a minimum time constant for a given background level and a constant D*-bandwidth product for background limited extrinsic photoconductivity, as has been predicted. When the load resistance is smaller than that of the detector, the bandwidth can be extended, and the D*-bandwidth product increased provided that background limited operation can be maintained with the smaller load resistance.

Escher, J. S., Antypas, G. A. and Edgecumbe, J.

HIGH-QUANTUM-EFFICIENCY PHOTOEMISSION FROM AN InGaAsP PHOTOCATHODE

An improved InGaAsP quaternary III-V material has been developed for near-ir photocathode applications. A quantum efdficiency of 9% per incident photon at 1.06 μm from a 1.15 eV band-gap sample has been achieved at room temperature.

Ewing, W. S., Shepherd, F. D., Capps, R. W. and Dereniak, E. L.

APPLICATIONS OF AN INFRARED CHARGE-COUPLED DEVICE SCHOTTKY DIODE ARRAY
IN ASTRONOMICAL INSTRUMENTATION

An infrared staring mode imaging sensor was designed to operate
with the NASA Infrared Telescope Facility (IRTF) at Mauna Kea, Hawaii.
The sensor has a 0.5 arc sec spatial resolution over a 10 arc sec
field and is matched to the f/35 focal ratio of the telescope. The
focal plane device is a back-side illuminated Schottky infrared
charge-coupled device (IRCCD) mosaic with a photoresponse from 1 to
5 μm. Astronomical observations are reported at 2.2 μm where the
IRCCD quantum efficiency is substantially greater than values normally
reported for the 3.4 to 4.2 μm thermal imaging band.

Reprinted with permission from Optical Engineering, Vol. 22, No. 3,
pp 334 (1983).

Fiorito, G., Gasparrini, G. and Svelto, F.

Hg-IMPLANTED $Hg_{1-x}Cd_xTe$ INFRARED PHOTOVOLTAIC DETECTORS IN THE 8- TO
14- μm RANGE

p-n junction photovoltaic detectors have been obtained by Hg
implantation in $Hg_{1-x}Cd_xTe$ for the 2- to 14- μm band. High-
sensitivity photodiodes were achieved over the entire range quoted;
the most interesting results are those concerning the 8- to 14- μm
range; peak detectivity better than 10^{10} cm $Hz^{\frac{1}{2}}$ W^{-1}, a zero-bias
resistance area product between 2.4-0.12 $\Omega\,cm^2$, specific responsivity
between 1.2 x 10^4 and 5 x 10^2 V W^{-1}, and quantum efficiency between
65% and 45% were obtained.

Reprinted with permission from Applied Physics Letters, Vol. 23, No.
8, pp 448 (1973).

Fiorito, G., Gasparrini, G. and Svelto, F.

ADVANCES IN Hg IMPLANTED $Hg_{1-x}Cd_xTe$ PHOTOVOLTAIC DETECTORS

Some recent results on Hg implanted $Hg_{1-x}Cd_xTe$ photovoltaic
detectors are reported. At 77K, peak detectivities equal to 2.8 x
10^{11}, 5.9 x 10^{10} and 4 x 10^{10} cm $Hz^{\frac{1}{2}}$ W^{-1} with quantum efficiencies
exceeding 90% were measured for selected diodes respectively at 3.7, 8
and 10.1 μm in reduced background conditions (i.e. FOV = 60°). Zero
bias resistance area products as high as $10^4 \Omega\,cm^2$ at 5 μm and 24 $\Omega\,cm^2$
cm^2 at 9.5 μm were observed for 1.2 x 10^{-3} cm^2 detectors. The
capacitance-voltage measurements indicate an abrupt junction; at
strong reverse bias an anomalous effect appears. The response time of

diodes with 3×10^{-4} cm^2 area is 1 nsec.

Reprinted with permission from Infrared Physics, Vol. 15, pp 287.
Copyright (1975), Pergamon Press.

Fiorito, G., Gasparrini, G, and Svelto, F.

PROPERTIES OF Hg IMPLANTED Hg$_{1-x}$Cd$_x$Te INFRARED DETECTORS

Experimental performance parameters of Hg implanted Hg$_{1-x}$Cd$_x$Te photovoltaic detectors are analyzed. At 77K, for 8-14 μm band, a comparison is made between performance and theoretical ultimate diffusion limits in low frequency direct detection. Experimental features are well-explained by a model based on the Auger band-to-band process for carrier recombination. Peak detectivities exceeding 10^{11} cm Hz$^{\frac{1}{2}}$ W^{-1}, external quantum efficiencies as high as 90%, and zero-bias resistance-area products better than 1 Ωcm^2 have been achieved in devices with 12 μm cutoff wavelengths. In the 3-5 μm band performances are far from the diffusion limit. Notwithstanding, at 77K zero-bias resistance-area products are better than 10^4 Ωcm^2 and detectivities of the order of 10^{12} cm Hz$^{\frac{1}{2}}$ W^{-1} were observed at 5 μm.

Predominant generation-recombination contribution are present at room temperature in 1- 1.3 μm photodiodes whose detectivities, primarily limited by the Johnson noise, at 1.3 μm are higher than 10^{11} cm Hz$^{\frac{1}{2}}$ W^{-1} at 300 K.

The high frequency response of the photodiodes is also discussed. Response times as low as 0.5 ns are reached despite some limitations arising from the implanted layer sheet resistance.

Reprinted from Applied Physics, Vol. 17, pp 105, (1978).

Fisher, D. G., Enstrom, R. E., Escher, J. S. and Williams, B. F.

PHOTOELECTRON SURFACE ESCAPE PROBABILITY OF (Ga,In)As:Cs-O IN THE 0.9 TO 1.6 μm RANGE

Ga$_{1-x}$In$_x$As alloys in the composition range $0 \leqslant x \geqslant 0.52$ and band-gap (E$_g$) range of 1.38 to 0.74 eV were activated with Cs and O$_2$. Samples of different carrier concentrations were investigated. For band-gaps down to about 0.8 eV, the photothreshold was equal to the band-gap. The longest wavelength threshold determined was 1.58 μm. To the best of our knowledge, this represents the longest wavelength response yet achieved for photoemission into vacuum from a III-V compound. The surface escape probability, B, was derived from the quantum yield data for each sample. The B-vs-E$_g$ data were analyzed according to a surface escape model which includes the effects of (i) a finite-width initial energy distribution of photoexcited carriers, (ii) the bent-band region and (iii) various types of surface potential barriers. Surface escape probability data pertaining to a single doping density could be explained by a model that includes only a work-function barrier or simple step potential. However, in order to explain the data for the several doping concentrations in a consistent manner, it was necessary to include an electron-semitransparent energy barrier above the vacuum level. A barrier width of 8 A gives good agreement with the experimental data. This dimension is consistent with the thickness of the Cs-O activation layer which was experimentally determined to be on the order of a monolayer. These results are interpreted in terms of a surface double-dipole model.

Reprinted with permission from Journal Applied Physics, Vol. 43, No. 9, pp 3815, (1972).

Foyt, A. G., Lindley, W. T. and Donnelly, J. P.

n-p JUNCTION PHOTODETECTORS IN InSb FABRICATED BY PROTON BOMBARDMENT

We have fabricated n-p junction photovoltaic detectors in InSb using proton bombardment to create the n-type layer. At 77°K, diodes which were 20 mils in diameter had zero-bias resistances of several hundred thousand ohms. The peak detectivity at 4.9 μm of these diodes with a 2π, 300°K background was typically greater than 3×10^{10} cm $Hz^{\frac{1}{2}}$ /W with the largest value observed being 10^{11} cm $Hz^{\frac{1}{2}}$ /W. Diode quantum efficiencies near 35% were observed.

Reprinted with permission from Applied Physics Letters, Vol. 16, No. 9, pp 335, (1970).

Foyt, A. G., Harman, T. C. and Donnelly, J. P.

TYPE CONVERSION AND n-p JUNCTION FORMATION IN $Hg_{1-x}Cd_xTe$ PRODUCED BY PROTON BOMBARDMENT

n-p junction photovoltaic detectors in $Hg_{1-x}Cd_xTe$ with x = 0.50, 0.31, and 0.25 using proton bombardment to create the n-type layer have been fabricated. Peak detection sensitivities were in the wavelength range 1.6-6 μm. Although high-sensitivity photodiodes were obtained with each composition, the best results were obtained with the x = 0.31 material, which has a peak response at 3.8 μm. At 77°K, 15 mil x 15 mil diodes made with this material had zero-bias impedances of several megaohms. The peak detectivity at 3.8 μmwas 9 x 10^{11} cm $Hz^{\frac{1}{2}}$/W in reduced background and the quantum efficiency at the peak was 29%.

Reprinted with permission from Applied Physics Letters, Vol. 18, No.

8, pp 321, (1971).

Gallinaro, G. and Varone, R.

CONSTRUCTION AND CALIBRATION OF A FAST SUPERCONDUCTING BOLOMETER

The construction and calibration of a fast superconducting bolometer, using a tin film evaporated onto a thin Al_2O_3 substrate is described. The circuit used for calibration and a typical response curve are illustrated. The results show that a responsivity of 1400 VW^{-1} at a frequency of 12.5 KHz, a time constant of 3 μsec and an NEP of 10^{-13} $Hz^{\frac{1}{2}}$ have been achieved. From this a D* of 3×10^{12} cm $Hz^{\frac{1}{2}}$ W^{-1} is calculated.

Cryogenics, Vol. 15, pp 292, (1975).

Ghosh, C.

PHOTOEMISSIVE MATERIALS

Photoemissive materials usually have a quantum efficiency of greater than 10^{-2} electrons per incident photon, all materials in this category are semiconductors. The mechanism of photoemission is discussed, and several examples of the more popular photoemissive materials are presented with their spectral response curves.

Negative electron affinity photocathodes, such as Si and GaAs have high quantum efficiencies and respond out to approximately 1.1 μm. To extend this response further into the infrared region of the spectrum transferred electron photoemitters have been proposed.

Proceedings of SPIE, Vol. 346, pp 62, (1982).

Groves, S. H.

TEMPERATURE-GRADIENT LPE GROWTH OF $Pb_{1-x}Sn_xTe$

An apparatus has been constructed for liquid-phase epitaxy that permits the application of a strong temperature gradient normal to the substrate-solution interface with minimal, unwanted gradients in other directions. A morphology problem with growth of $Pb_{1-x}Sn_xTe$ alloys on PbTe substrates slightly misoriented from the (100) plane has been used to test the induced temperature gradient growth. No improvement is found over growth with a conventional LPE slider, and very large gradients cause a deterioration in morphology and growth stability. A small systematic variation of the Sn/Pb ratio in the grown layers with rate of cooling is found, but no variation with temperature gradient can be detected. Thus, contrary to our conjecture made before this study, the presence or absence of an induced temperature gradient cannot be used to explain the systematic differences between the data of different workers for the liquid-solid tie-line behaviour of the Pb-Sn-Te system.

Reprinted with permission from Journal of Electronic Materials, Vol. 6, No. 2, pp 195 (1977), a publication of the Metallurgical Society of AIME, Warrendale, Pennsylvania.

Hadni, A., Thomas, R., Mangin, J. and Bagard, M.

PYROELECTRIC DETECTORS FOR SUBMILLIMETRE WAVES

To increase the detectivity we have worked in two ways: the reduction of crystal thickness: and the use of new materials and low temperatures. Pyroelectric data down to 1.5 K have been obtained for TGS, TGSe, KDP, thiourea, Li_2SO_4 H_2O and GASH.

Reprinted with permission from Infrared Physics, Vol. 18, pp 663, copyright (1978) Pergamon Press.

Hamilton, C. A., Phelan, Jr., R. J. and Day, G. W.

PYROELECTRIC RADIOMETERS

Pyroelectric detectors can be used in radiometric measurements with the benefit of a broad spectral response as typifies thermal devices, plus speed and sensitivity. A detector has been constructed in which the sensitive area acts as both an optical absorber and an electrical heater. Using this technique the problems of responsivity drift, temperature coefficients and optical calibration found in conventional radiometers are all eliminated. Four systems have been constructed at NBS in Boulder, Colorado and are undergoing evaluation. The minimum resolvable power is 10^{-7} watts in a ten second averaging time.

Optical Spectra, Vol. 9, No. 10, pp 37, (1975).

Harman, T. C.

CONTROL OF IMPERFECTIONS IN CRYSTALS OF $Pb_{1-x}Sn_xTe$, $Pb_{1-x}Sn_xSe$, AND $PbS_{1-x}Se_x$

Some phase diagrams of the pseudobinary alloy systems $Pb_{1-x}Sn_xTe$, $Pb_{1-x}Sn_xSe$ and $PbS_{1-x}Se_x$ relevant to crystal growth and atomic imperfections are presented. Two useful crystal growth techniques are described. Information pertaining to such macroscopic imperfections as metallic inclusions, low-angle gain boundaries, and dislocation etch pits are discussed. Some experimental data relating lattice point defects at the extremes of the alloy existence fields to temperature will be presented for the above three alloys. The effects of various foreign impurities on rock-salt structure IV-VI compounds and alloys are reviewed. A technique for removal of some of the foreign impurities in PbTe and SnTe will be given along with mass spectrographic analyses. Finally, the metallurgical characterization of $Pb_{1-x}Sn_xTe$ crystals with respect to imperfections using etching techniques will be summarized.

Reprinted with permission from The Journal of Nonmetals, Vol. 1, pp 183, (1973).

Hauser, M. G. and Notarys, H. A.

COMPOSITE BOLOMETERS FOR MILLIMETRE WAVE ASTRONOMY

Composite bolometers combining a short time constant with high detectivity at submillimeter and millimeter wavelengths have been developed. A 2mm x 2mm bolometer with a 5 msec time constant and a far infrared radiation NEP of 5×10^{-14} W $Hz^{-\frac{1}{2}}$ at 1.8 K has been produced.

Bulletin of The American Astronomical Society, Vol. 7, pp 409, (1975).

Hodges, D. T. and McColl. M.

EXTENSION OF THE SCHOTTKY BARRIER DETECTOR TO 70 μm (4.3 THz) USING SUBMICRON-DIMENSIONAL CONTACTS

Schottky barrier diode detection in both video and mixing modes of operation has been extended to 4.252 THz (70.5 μm) using 0.5 μm-diam diodes fabricated from heavily doped nonepitaxial n-type GaAs. These ultra small, and consequently ultralow capacitance, junctions were prepared using electron-beam lithography and have yielded the smallest reported series-resistance junction-capacitance product for a Schottky barrier diode. The advantages of nonepitaxial versus epitaxial Schottky barrier diodes for high-frequency operation is discussed.

Reprinted with permission from Applied Physics Letters, Vol. 30, No. 1, pp 5, (1977).

Hohnke, D. K. and Holloway, H.

EPITAXIAL PbSe SCHOTTKY-BARRIER DIODES FOR INFRARED DETECTION

The photovoltaic properties and infrared response of epitaxial PbSe Schottky-barrier diodes are reported. The temperature dependence of zero-bias resistance suggests that generation/recombination is the dominant transport mechanism. At $77^{o}K$ the devices are limited by the $290^{o}K$ background at f/0.6. With further reduction of the background they attain Johnson-noise-limited peak detectivities of 5×10^{11} cm $Hz^{\frac{1}{2}} W^{-1}$. The diode quantum efficiencies are reflection loss limited and reach 70%.

Reprinted with permission from Applied Physics Letters, Vol. 24, No. 12, pp 633, (1974).

Igras, E., Piotrowski, J. and Zimnoch-Higersberger, I.

INVESTIGATION OF ION IMPLANTED GRADED GAP (CdHg)Te PHOTODIODES

A range of donor ions and acceptor ions have been implanted into epitaxial layers of p and n type cadmium mercury telluride. The implantation was carried out at room temperature with doses of 10^{13}- 10^{15} ions cm^{-2} and at energies of 30-140 KeV. The samples were not subjected to any post implantation annealing process. In the case of donor implants an n-type layer was formed on p-type material. However, in the case of acceptor implants into n-type layers no formation of p-type layers was observed. The junctions were investigated using I-V and C-V characteristics and photoelectric properties. The device performance was strongly dependent on the p-type substrate preparation and only weakly on species, dose and energy of the implanted donor atom. High performance infrared photovoltaic devices were produced.

Electron Technology, Vol. 10, No. 4, pp 63, (1977).

James, L. W., Antypas, G. A., Uebbing, J. J., Yep, T. O. and Bell, R.

OPTIMIZATION OF THE $InAs_xP_{1-x}-Cs_2O$ PHOTOCATHODE

Zinc-doped InAsP liquid epitaxial layers with band-gaps between 0.4 and 1.34 eV were grown on InAs and InP substrates. The grown layers were 2-4 μm thick with mirror-smooth as grown surfaces. Preliminary phase diagram calculations based on Darken's quadratic formalism to describe the ternary liquid in equilibrium with the pseudobinary solid are in good agreement with the band-gaps of the grown layers determined by photoluminescense. The $InAs_xP_{1-x}-Cs_2O$ heterojunction barrier height as a function of composition has been measured using photoemission. For InAs the barrier is at 1.24 eV, and it decreases with decreasing arsenic concentration to a value of 1.16 eV for InAsP with a 1.27-eV band-gap. For $InAs_xP_{1-x}$ samples with band-gaps in the range 1.17-1.34 eV, high escape probabilities and efficient photoemission were observed. A typical cleaned (not cleaved) sample with a band-gap of 1.19 eV has a sensitivity of 600 μA/lm, 70 μA with a lumen source through a 2540 ir filter, a quantum efficiency of 1.5% at 1.06 μm, and a Γ escape probability of 0.08. This is the most sensitive infrared photocathode yet produced. All processing steps seem compatible with tube production. The effects of the heterojunction barrier are clearly visible with this material. The escape probability drops by an order-of-magnitude when the $InAs_xP_{1-x}$ band-gap is reduced to 0.05 eV below the barrier.

Reprinted with permission from Journal of Applied Physics, Vol. 42, No. 2, pp 580, (1971).

James, L. W., Antypas, G. A., Edgecumbe, J., Moon, R. L. and Bell, R.

DEPENDENCE ON CRYSTALLINE FACE OF THE BAND BENDING IN Cs_2O-ACTIVATED GaAs

Electron energy loss in the band-bending region of the p-type III-V semiconductor in a III-V photocathode is an important factor in determining the escape probability and the optimum doping. From measurements of photoelectric yield near threshold from Cs_2O-activated n-type GaAs, the position of the Fermi level at the GaAs-Cs_2O interface was determined for [110], [100], [111A] and [111B] surfaces. Assuming the Fermi-level position at the GaAs surface to be independent of doping, the band bending for p-type GaAs is greatest for the [111A] face and least for the [111B] face. The measured escape probabilities of photoexcited electrons from different crystalline faces of optimally activated $5 \times 10^{18}/cm^3$ Zn-doped liquid epitaxial GaAs correlate well with the band-bending measurements. The [111B] sample has an escape probability of 0.489 and a luminous sensitivity of 1837 μA/lm.

Reprinted with permission from Journal of Applied Physics, Vol. 42, No. 12, pp 4976, (1971).

Johnson, E. S. and Schmit, J. L.

DOPING PROPERTIES OF SELECTED IMPURITIES IN $Hg_{1-x}Cd_xTe$

The doping properties of selected impurities in $Hg_{1-x}Cd_xTe$ have been determined. Primary emphasis is on elements from Groups IB and IIIA, expected to substitute on the metal sublattice, and on elements from Groups VA and VIIA, expected to substitute on the Te sublattice. In addition, the behaviour of some elements from Group IV as well as behaviour of Li has been determined. Impurities were introduced into $Hg_{1-x}Cd_xTe$ either by diffusion or during crystal growth. Cu, Ag and Li are fast diffusing acceptors, Ga is a fast diffusing donor, Al and

Si are donors which require high diffusion temperature to effect
diffusion, P and As are slowly diffusing acceptors and Br is a slowly
diffusing donor. Sn appears to be inactive. In general, impurities
substituted on the metal sublattice are rapid diffusers while those
substituted on Te sites are slow diffusers.

Reprinted with permission from Journal of Electronic Materials, Vol.
6, No. 1, pp 25, (1977), a publication of The Metallurgical Society of
AIME, Warrendale, Pennsylvania.

Kinch. M. A.

COMPENSATED SILICON-IMPURITY CONDUCTION BOLOMETER

The capability and performance of compensated Si (Sb~2 x 10^{18} cm^{-3}
B~2 x 10^{17} cm^{-3}) impurity conduction bolometers as extremely sensitive
detectors of far-infrared radiation (2 mm to 30 μm) is described.
Electrical and far-infrared measurements indicate a NEP ~2.5 x 10^{-14}
W/Hz$^{\frac{1}{2}}$, and a response time ~10^{-2} sec, when operated at 1.5°K.

Reprinted with permission from Journal of Applied Physics, Vol. 42,
No. 13, pp 5861, (1971).

Kinch, M. A., Borrello, S. R., Breazeale, B. A. and Simmons A.

GEOMETRICAL ENHANCEMENT OF HgCdTe PHOTOCONDUCTIVE DETECTORS

Ohmic contacts pose a severe lifetime degradation problem for
photoconductive detectors whose length is comparable to a minority
carrier diffusion length. A photoconductive device is described which
greatly reduces the effect of ohmic contacts on effective
photoconductive lifetime. The theory of operation is presented
together with experimental data indicating significant enhancement in
photoconductive responsivity, effective minority carrier lifetime, and
1/f noise performance in n-type HgCdTe photoconductors.

Reprinted with permission from Infrared Physics, Vol. 17, pp 137.
Copyright (1977), Pergamon Press.

Kohn, E. S.

INFRARED IMAGING WITH MONOLITHIC, CCD-ADDRESSED SCHOTTKY-BARRIER
DETECTOR ARRAYS: THEORETICAL AND EXPERIMENTAL RESULTS

The theoretical basis for infrared imaging in the 3 to 5 μm
spectral band with CCD addressed silicon, Schottky-barrier mosaics is
presented. A unique approach is used which allows readout of majority

carrier signals with depletion mode CCD's. Photo-response, contrast, and noise relationships for this type of all solid-state sensor are derived. It is seen that the use of the Schottky-barrier, internal-photoemission process, which is independent of lifetime and doping variations in the silicon wafer, leads to at least a factor of 100 improvement in infrared photo-response uniformity. This advance permits for the first time the development of infrared cameras that are not limited by fixed pattern noise. Systems considerations such as cooling requirements, noise mechanisms, and cut off wavelengths, are related to signal contrast and noise-equivalent-temperature (NEΔT). A charge-coupled imager sensitive to infrared light as far out as 3.5 μm has been fabricated and operated. It consists of a linear array of 64 Pd:p-Si Schottky-barrier detectors adjacent to a three-phase charge-coupled shift register. A single transmission gate, when pulsed on, coupled each detector to its associated shift register gate, thus reverse-biasing the detectors. The charges transferred to the shift register are then read aout sequentially to produce the video signal. It is demonstrated that in this mode of operation, the IR-CCD is particularly immune to non-uniformities in substrate doping and in MOSFET pinch-off voltage. Visible images were sensed directly by illumination of the shift register through the gaps as well as through the unthinned substrate. Infrared images (1.1 μm $<\lambda<$ 3.5 μm) were sensed by the Schottky-barrier detectors illuminated through the (transparent) substrate. The two imaging modes could be easily distinguished by their spectral sensitivities as well as by their responses to changes in their separate integration times. All IR measurements were made at 77°K. Uniformity was within a few percent, and objects at 110°C could be detected. A scheme for observing low-contrast, thermal scenes without requiring the charge-coupled shift register to carry the entire background signal has been implemented in the design of this chip. Operation in this mode was also demonstrated.

Reprinted from Proceedings of the International Conference on Applications of CCDs (San Diego) (1975), pp 59.

Kosonocky, W. F., Elabd, H., Erhardt, H. G., Shallcross, F. V., Meray G. M., Villani, T. S., Groppe, J. V., Miller, R. and Frantz, V. L.

DESIGN AND PERFORMANCE OF 64 X 128-ELEMENT PtSi Schottky-Barrier IR-CCD FOCAL PLANE ARRAY

A 64 x 128-element IR-CCD focal plane array was developed with high-performance PtSi Schottky-barrier detectors. The buried-channel CCD imager has an interline transfer organization with 22% detector area efficiency and 120 x 60 μm pixel size. The high-performance "thin" platinum silicide detectors have cut-off wavelength of about 6.0 μm and quantum efficiency of 4.0 to 1.0% in the 3.0 to 4.5 μm spectral range. Depending on the processing parameters and operating voltage, the measured dark current density of these IR detectors at 77K is in the range of 5 to 60 nA/cm^2. High quality thermal imaging has been obtained with the 64 x 128-element PtSi Schottky-barrier IR-CCD focal plane array in a TV compatible IR camera. The IR-CCD camera operates at 60 frames per second without vertical interlacing.

Krall, H. R., Helvy, F. A. and Persyk, D. E.

RECENT DEVELOPMENTS IN GaP(Cs)-DYNODE PHOTOMULTIPLIERS

Recent developments in GaP(Cs)-dynode photomultipliers are reviewed. The physics of secondary emission is discussed and models are described which explain secondary emission from conventiuonal dynode materials and from GaP(Cs) dynodes. The concept of negative electron affinity is explained in terms of an energy diagram. Device performance is reported with regard to both electron and time resolution. A general discussion of the characteristics of GaP(Cs)-dynode photomultipliers is included.

Kruse, P. W.

THE PHOTON DETECTION PROCESS

The performance of optical and infrared detectors can be described in terms of unique figures of merit which enable the performance of a system to be predicted and evaluated. These devices can be categorized as photon, thermal or wave interactive detectors and these mechanisms and the fundamental electrical noise in the detector are discussed in detail.

A review of the various figures of merit and the fundamental limits on performance are given, and finally a summary of the results obtained from state of the art optical and infrared detectors is presented.

Kutzscher, E. W.

INFRARED AND ULTRAVIOLET TECHNIQUES: EVOLUTION OF INFRARED TECHNOLOGY

Infrared radiation emitted or absorbed and reflected by solids, liquids or gases can be detected and analysed by the use of suitable detectors. The selective infrared spectrum of gases and liquids is typical for the chemical composition of the material and is the basis of infrared spectroscopy. Theoretical and experimental research and the development of all necessary instrument components has resulted in successful applications of infrared spectroscopy in science and technology. The development of thermal and photon detectors for infrared radiation constitutes an important basis for all technical applications of infrared in both fields, spectroscopy and also detection, position-finding and measurement of infrared radiators. Systematic research in many special fields has resulted in successful development of infrared instrumentation for military and civilian applications. These fields include: radiation characteristics of emitters of interest as well as their surroundings and backgrounds, atmospheric attenuation of radiation energy, infrared optics, scanning techniques and detectors.

Reprinted with permission from Laser '75. Optoelectronics Conference Proceedings, pp 206 (1975).

Lock, P. J.

DOPED TRIGLYCINE SULFATE FOR PYROELECTRIC APPLICATIONS

Crystals of triglycine sulfate (TGS) grown from solutions containing alanine are found to give biased ferroelectric hysteresis loops. The material is therefore always in a fully poled state when below its Curie point in the absence of large external fields. The values of e' and e" are lower than for pure TGS. Pyroelectric infrared detectors using this material approach within an order of magnitude of the limit of sensitivity for uncooled thermal detectors.

Reprinted with permission from Applied Physics Letters Vol. 19, No. 10, pp 390, (1971).

Long, D. and Schmit, J. L.

MERCURY CADMIUM TELLURIDE AND CLOSELY RELATED ALLOYS

This chapter concentrates almost entirely on the alloy cadmium mercury telluride.

The basic material properties and the important aspects of intrinsic infrared detector theory applicable to these materials is presented. The methods used to prepare high quality single crystals of the alloy or the fabrication and properties of typical detectors are described. The conclusion summarizes the present status of the devices covered in this chapter and indicates the prospects for their future development.

Semiconductors and Semimetals Vol. 5, Infrared Detectors, pp 175, (1970), Editors R K. Willardson and A. C. Beer, Academic Press.

Longo, J. T., Cheung, D. T., Andrews, A. M., Wang, C. C. and Tracy, J.

INFRARED FOCAL PLANES IN INTRINSIC SEMICONDUCTORS

In this paper, the state-of-the-art of intrinsic semiconductor detector arrays and projected future areas of development are reviewed. Infrared focal planes in instrinsic semiconductors offer advantages over extrinsic semiconductor structures in both operating temperature and quantum efficiency. Although the device function of spectral filtering and detection of the incident photon flux is now well understood in intrinsic semiconductors, the function of signal processing has only recently been investigated. As a result, research is directed toward implementation of both hybrid devices, in which the signal processing is accomplished in a silicon multiplexer which is physically and electrically interfaced with an intrinsic semiconductor detector array, and monolithic charge transfer devices in which detection and signal processing are accomplished in the same semiconductor. In the monolithic approach, charge transfer devices have been demonstrated in InSb, and it is likely that similar devices will be realised in InSb related alloys and HgCdTe in the near future. Demonstration of a non-MIS charge transfer design would open up the monolithic approach to the IV-VI alloys. Hybrid focal planes incorporating ∼1000 element photodiode arrays have been realized in the III-V and the IV-VI alloys; the detector-multiplexer interface circuit will remain one of the key technical issues in the achievement of a high-performance hybrid focal plane.

Marine, J. and Motte, C.

INFRARED PHOTOVOLTAIC DETECTORS FROM ION-IMPLANTED $Cd_xHg_{1-x}Te$

n-p photovoltaic detectors in $Cd_xHg_{1-x}Te$ using aluminium implantation to create the n-type region have been fabricated. Implanted diodes made in this material with a composition x = 0.18, which corresponds to a band gap of 0.1 eV, had a zero-bias resistance of 1 KΩ at 77°K for a 200 x 250 μm sensitive area. The spectral response was almost flat in the range 4-12 μm with a quantum efficiency as high as 57% at 10.6 μm. The measured detectivity for a

$30°$ field of view is 7.3×10^{10} cm $Hz^{\frac{1}{2}}$ W^{-1} at 10.6 μm. This is one of the highest detectivities ever reported at this wavelength. Finally, the frequency response of these diodes used in a heterodyne detection system operating with a CO_2 laser was 1 GHz.

Reprinted with permission from Applied Physics Letters, Vol. 23, No. 8, pp 450, (1973).

Martinelli, R. U.

INFRARED PHOTOEMISSION FROM SILICON

Infrared photoemission from p-type silicon has been observed with a threshold of 1.1 eV. The surface of the sample has been activated to a state of effective negative electron affinity. The escape depth for thermal photoelectrons is 5.5 μ (microns) and surface escape probability is 0.18.

Reprinted with permission from Applied Physics Letters, Vol. 16, No. 7, pp 261, (1970).

Martinelli, R. U. and Fisher, D. G.

THE APPLICATION OF SEMICONDUCTORS WITH NEGATIVE ELECTRON AFFINITY SURFACES TO ELECTRON EMISSION DEVICES

Semiconductors with negative electron affinity (NEA) surfaces are used as photoemitters, secondary emitters, and cold-cathode emitters. A comprehensive review of the characteristics and applications of these materials is presented, the concept of NEA is described, and a comparison is made between NEA and conventional emitters. Electron generation, transport, and emission processes of NEA emitters are discussed. NEA III-V compound photocathodes, especially GaAs, are described with respect to their fabrication, performance, and applications to photomultipliers and image intensifier tubes. The structure and performance of NEA secondary emitters are presented. NEA GaP secondary-emission dynodes represent the most important device application. NEA cold-cathodes, using GaAs, Ga(As, P), or Si, have been investigated, and their performance characteristics are summarized. NEA Si cold-cathodes have been incorporated in developmental TV camera tubes. The characteristics of these tubes are reviewed.

Copyright © 1974 IEEE. Reprinted with permission from Proceedings of the IEEE, Vol 62, No. 10, pp 1339, (1974).

McColl, M., Hodges, D. T. and Garber, W. A.

SUBMILLIMETER-WAVE DETECTION WITH SUBMICRON-SIZE SCHOTTKY-BARRIER
DIODES

Schottky-barrier diode detection has been extended to 7.2 THz
(42 μm) using 0.5- μm-diam diodes. The diodes were fabricated on
bulk-doped n-type GaAs using electron lithographic techniques;
diameters as small as 1000 A have been achieved. A new approach in
Schottky-barrier design, the contact array diode, is proposed. The
diode is fabricated from readily available bulk doped material, and a
performance is indicated that is competitive to the conventional
epitaxial Schottky-barrier mixer well into the submillimeter
wavelength region. A scanning electron microscope (SEM) photograph of
diode array structure is shown.

McNally, P. J.

ION IMPLANTATION IN InAs and InSb

Ion implantation in InAs and InSb with sulfur and zinc ions has
been used to fabricate p-n junction diodes which have been
characterized for their infrared detector properties. Planar mosaic
infrared detectors have been produced in both materials with good
characteristics. Blackbody detectivities (D^*_{BB}) of 2×10^9 cm $Hz^{\frac{1}{2}} W^{-1}$
and 4 per cent uniformity have been measured for InAs. Similarly,
InSb has shown D^*_{BB} values of 1.3×10^{10} cm $Hz^{\frac{1}{2}} W^{-1}$ and 49 per cent
uniformity between elements in an array.

Experimental range-energy data for zinc in InSb has been obtained
between 0.2 and 1.8 MeV and compared with predicted values from LSS
theory. Theory predicts a deeper range than experimental values
indicate; however, the differences are sufficiently small to make the
curves useful for device design. The slopes of the curves indicate
that a large component of nuclear stopping predominates in this energy
region.

Melngailis, I. and Harman, T. C.

SINGLE CRYSTAL LEAD TIN CHALCOGENIDES

The semiconductor alloys $Pb_{1-x} Sn_x Te$ and $Pb_{1-y} Sn_y Se$ have

composition dependent energy gaps which can be made arbitrarily
small. This special property together with the ease in producing
single crystals with excellent homogeneity and good quality has made
these alloys particularly useful for long wavelength infrared
detectors as well as lasers. This chapter reviews the present
understanding of the band structure of these alloys, the materials
properties and crystal preparation techniques, and the fabrication and
properties of photovoltaic and photoconductive devices.

Semiconductors and Semimetals Vol. 5, Infrared Detectors, pp 111,
(1970), Editors R. K. Willardson and A. C. Beer, Academic Press.

Moss, T. S., Burrell, G. J. and Ellis, B.

SEMICONDUCTOR OPTO-ELECTRONICS

The term opto-electronics covers the basic physical phenomena and
device behaviour which arise from the interaction between electromagnetic radiation and the electrons in a solid. The first nine
chapters of this book are devoted to theoretical topics, covering the
interaction of electromagnetic waves with solids, dispersion theory
and absorption processes, magneto-optical effects, and non-linear
phenomena. Theories of photo-effects and photo-detectors are treated
in detail, as are the theories of radiation generation and the
behaviour of semiconductor lasers and lamps. The discussion on the
wide range of optoelectronic materials has been limited to the group
IV elements, the III-V compounds and a selection of the most important
chalcogenides. This includes virtually all materials of current
applications interest as well as the more important optoelectronic
semiconductors.

Semiconductors and Optoelectronics (1973), Butterworths.

Moustakas, T. D. and Connell, G. A. N.

AMORPHOUS Ge_xH_{1-x} BOLOMETERS

Selected materials from the amorphous Ge_xH_{1-x} system are evaluated
at 300 K as thermistor bolometers operating at photon energies between
approximately 0.2 and 2.5 eV. The devices are deposited on glass and
sapphire substrates by rf sputtering of polycrystalline Ge in an
argon-hydrogen atmosphere. Their current voltage characteristics,
their response times, the frequency and spectral dependence of their
responsivities, and the frequency dependence of the noise in a 3-Hz
bandwidth are all measured. Response times between 1 and 10 msec and
$D_\lambda*(0.633, 13, 1)$'s of between 5×10^6 and 5×10^7 $W^{-1}Hz^{\frac{1}{2}}$ cm are
obtained, using a calibrated He-Ne laser ($\lambda = 0.633$ μm) as an
excitation source. (At this wavelength, all nonreflected incident

light is absorbed).

Reprinted with permission from Journal of Applied Physics, Vol. 47, No. 4, pp 1322, (1976).

Nayar, P. S.

A NEW FAR INFRARED DETECTOR

We have constructed a far infrared bolometer detector by using a single crystal chip of p-type undoped thallium selenide as the temperature sensitive element. The bolometer operates at 1.6 K and achieves a noise equivalent power of 8.3×10^{-15} W/ Hz$^{\frac{1}{2}}$ with a time constant of 22 msec. The measured responsivity is 6×10^{5} V/W. The detector should be suitable for infrared and microwave applications.

Reprinted with permission from Infrared Physics, Vol. 14, pp 31, copyright (1974), Pergamon Press.

Nudelman, S.

IMAGE INTENSIFICATION DEVICES AND APPLICATIONS

Image intensifiers are used to detect photons and energetic particles and to provide an output by wavelength conversion and/or intensification.

Luminescent surfaces achieve intensification when a high energy input is multiplied by the conversion to and emission of low energy, visible photons. Vacuum tube and solid state intensifiers operate by the application of a potential difference through which electrons gain sufficient energy to generate cathodoluminescent or electroluminescent emission respectively. These devices and their components are reviewed and many examples of the imagery obtained are illustrated.

Electronic Imaging Conference, London 1978, pp 253, Academic Press.

Oba, K. and Rehak, P.

STUDIES OF HIGH-GAIN MICRO-CHANNEL PLATE PHOTOMULTIPLIERS

The characteristics and performance of several kinds of high-gain micro-channel plate photomultipliers have been investigated. Special attention was directed towards i) lifetime studies, ii) performance in the magnetic field, and iii) timing properties. Lifetime studies include separate investigations of the photocathode quantum

efficiency degradation caused by ion feedback, and the deterioration
of the micro-channel plate gain. The dependence of the micro-channel
plate photomultiplier gain on the intensity and the direction of the
magnetic field (up to 7 kGauss) is reported.

Olsen, G. H., Szostak, D. J., Zamerowski, T. J. and Ettenberg, M.

HIGH PERFORMANCE GaAs PHOTOCATHODES

Improved negative electron affinity (NEA) photoemission
sensitivity has been obtained from GaAs reflection photocathodes grown
from a stoichiometric (rather than arsenic-rich) vapor phase on (100)
substrates. Sensitivities as high as 2150 μA/lm with electron
diffusion lengths of ∼ 5 μm and escape probabilities of ∼0.55 are
reported.

Polla, D. L. and Sood, A. K.

SCHOTTKY BARRIER PHOTODIODES IN p $Hg_{1-x}Cd_xTe$

Schottky barrier photodiodes have been fabricated on p $Hg_{1-x}Cd_xTe$
($0.20 < x < 0.38$) with aluminium and chromium as barrier metals. Various
electrical characterizations have been carried out to determine
barrier heights and the results are found to be in excellent agreement
with Schottky thermionic emission theory. These photodiodes have
also been used to determine the minority carrier lifetime and
diffusion length in p $Hg_{1-x}Cd_xTe$.

Porter, S. G.

A BRIEF GUIDE TO PYROELECTRIC DETECTORS

The principles of operation of a pyroelectric detector are
summarized and responsivity, noise, noise equivalent power and
detectivity are derived. The relationship between responsivity and
frequency is discussed as are the various sources of noise.

There follows a discussion of the relative merits of the five principal pyroelectric materials in common use: Triglycine-Sulphate, Lithium Tantalate, Strontium Barium Niobate, Pyroelectric Ceramics and Polyvinylidene Fluoride. Factors influencing the choice of material are outlined and summarized in performance curves for three sizes of detector.

Mention is made of a variety of detector types currently available and some of the current applications. A very brief summary of possible future developments in included.

Reprinted with permission from Ferroelectrics, Vol. 33, pp 193, (1981).

Putley, E. H.

THE PYROELECTRIC DETECTOR

Many crystals exhibit spontaneous electric polarization, and for an insulator it is possible to produce a charge distribution at the surface. Thus if the temperature of the material is changed the dipole moment may also change and produce an observable external electric field, this is the pyroelectric effect and the exploitation of this to detect infrared radiation is discussed. To calculate the performance of a pyroelectric detector both the thermal and electrical characteristics and the detector's noise sources are evaluated, and expressions for the Noise Equivalent Power and the D* obtained. Detectors have been constructed and an NEP of approximately 10^{-9} W $Hz^{-\frac{1}{2}}$ at low frequencies, and response times of 100 nsec have been achieved.

Semiconductors and Semimetals, Vol. 5, Chapter 6, pp 259, (1970). Editors Willardson, R. K. and Beer, A. C., Academic Press.

Putley, E. H.

InSb SUBMILLIMETER PHOTOCONDUCTIVE DETECTORS

The characteristics of the indium antimonide submillimeter photoconductive detectors which are based on the physical mechanism of hot electron photosensitivity are reviewed. The design and construction of these detectors, and the performance including the responsivity, speed of response, NEP and D* are discussed.

These devices were first developed to meet the requirement for a fast detector in submillimeter plasma diagnostics, however, more recently they have found applications in radio and rocket-borne astronomy.

Semiconductors and Semimetals, Vol. 12, Infrared Detectors II, Chap.
3, pp 143, (1972). Editors, Willardson, R. K. and Beer, A. C.,
Academic Press.

Putley, E. H.

THERMAL DETECTORS

The general principles of thermal detectors and the evaluation of
their performance are discussed, and then the more important devices
including the thermopile, the bolometer, the Golay Cell and the
pyroelectric detector are described in detail. The application of
these sensors and in particular their use in infrared imaging systems
is also presented.

Topics in Applied Physics, Vol. 19, Optical and Infrared Detectors,
pp 71, (1977). Editor, Keyes, R. J., Springer Verlag.

Putley, E. H.

THE APPLICATIONS OF PYROELECTRIC DEVICES

After reviewing briefly the characteristics of pyroelectric
detectors (including the pyroelectric vidicon) the principal
applications of pyroelectric devices are discussed. These range from
high technology applications such as radiometers in weather
satellites and interplanetary probes to simple thermal sensors with
widespread commercial and industrial applications and make important
contributions to absolute radiometry, infrared spectroscopy and to
laser research. In addition to its use in simple thermal imaging
systems the pyroelectric vidicon is now widely used in laser
interferometers and other applications requiring a two dimensional
detector. Finally the possibility of future development of a two
dimensional pyroelectric/CCD array is considered.

Reprinted with permission from Ferroelectrics, Vol. 33, pp 207,
(1981), copyright © Controller HMSO, London.

Reine, M. B., Sood, A. K. and Tredwell, T. J.

PHOTOVOLTAIC INFRARED DETECTORS

The status of cadmium mercury telluride photovoltaic infrared
detector technology is reviewed, in particular p-n junction devices
and recent work on Schottky Barrier photodiodes. 3-5 μm devices
operating at 190 K and 8-12 μm devices at 77K have been fabricated by
ion implantation. For these detectors the I-V characteristics

generally dominated by the diffusion current in the p region, and work is now aimed at reducing this. At lower operating temperatures the conventional diffusion current is negligible and the devices are limited by leakage current mechanisms, in particular interband tunnelling. A summary of device results obtained by ion implantation, diffusion, electron irradiation, sputtering and pulsed laser irradiation are presented.

Semiconductors and Semimetals, Vol. 18, Chapter 6, pp 201, (1981). Editors, Willardson, R. K. and Beer, A. C., Academic Press.

Richards, P. L., Clarke, J., Hoffer, G. I., Nishioka, N. S., Woody, D. P., and Yeh, N. H.

COMPOSITE BOLOMETERS FOR SUBMILLIMETER WAVELENGTHS

Two types of ^{4}He temperature composite submillimeter wave bolometers have been developed, using metal film absorbing elements. Dark electrical noise equivalent powers of 2×10^{-15} W Hz$^{-\frac{1}{2}}$ and 3×10^{-14} W Hz$^{-\frac{1}{2}}$ have been measured in large area bolometers with superconducting and semiconducting thermometers, respectively. Experiments confirm the theoretically predicted frequency independent absorptivity of ~0.5 in a single pass. These results were obtained using a 135 μm thick sapphire substrate, if this was reduced to 30 μm significant improvement is expected.

Proceedings of the 2nd International Conference on Submillimeter Waves and their Applications, pp 64, (1976), IEEE.

Roberts, C. G.

HgCdTe CHARGE TRANSFER DEVICE FOCAL PLANES

Metal-Insulator-Semiconductor (MIS) detectors fabricated in HgCdTe offer significant advantages for focal plane applications. These detectors perform noise free signal integration directly in the HgCdTe, can be formed in either n- or p-type material, do not require formation of a metallurgical junction, and are easy to interface to low power signal processing integrated circuits in silicon. Furthermore, this device technology readily facilitates the formation of charge transfer devices which perform some signal processing in the HgCdTe prior to transfer of the signal to the silicon processor.

This paper discusses MIS detector design considerations (i.e., well capacity, dark current, operating temperature, and material requirements) for HgCdTe charge transfer device arrays suitable for MWIR (3-5 μm, and LWIR (8-10 μm) applications.

Several charge transfer device configurations have been developed.
These structures include charge coupled device (CCD) arrays, charge
injection device (CID) arrays, and a new IR device structure, the
charge imaging matrix (CIM) array. These devices are based on the
common MIS integrated circuit technology, but have performance
characteristics which make them particularly well suited for
different system applications. This paper discusses the state of
development, the major performance characteristics, and systems
applicability for each device.

Reprinted with permission from Proceedings SPIE, Vol. 443, pp 131,
(1983).

Rolls, W. H. and Eddolls, D. V.

HIGH DETECTIVITY $Pb_xSn_{1-x}Te$ PHOTOVOLTAIC DIODES

$Pb_xSn_{1-x}Te$ diodes have been made with detectivities up to 10^{11} cm
$Hz^{\frac{1}{2}} W^{-1}$ at 10.6 μm and 77K. Diodes with a peak response at 12 and
13 μm have detectivities greater than 50% of the 180° F.O.V.
background limited value. Measurements from 70 Hz to 50 kHz have
shown that 1/f noise is negligible if the diode is biased to the zero
volts point on the I-V characteristic. The same result will be
obtained by operating the diode into the d.c. short circuit input of a
virtual earth current amplifier. The advantages of this mode of
operation for thermal imaging applications and CO_2 laser detection are
discussed.

Reprinted with permission from Infrared Physics, Vol.13, pp 143,
copyright (1973), Pergamom Press.

Roundy, C. B., Byer, R. L., Phillion, D. W. and Kuizenga, D. J.

A 170 psec PYROELECTRIC DETECTOR

A $LiTaO_3$ pyroelectric detector with less than a 200 psec response
time has been designed and tested. The detector element is coated
with a flat spectral response fast thermal block and is impedance
matched into a 13 GHz bandwidth Tektronic S-4 sampling head. The
response was measured at 1.06 μm using a Q-switch mode-locked Nd:YAG
laser source with 100 psec pulses at 54 kW peak power. The measured
rise time is 170 $\pm$ 30% psec which is longer than the 50 psec RC time
constant for the element.

Reprinted with permission from Optics Communications, Vol. 10, No. 4,
pp 374, (1974).

Sclar, N.

EXTRINSIC SILICON DETECTORS FOR 3-5 AND 8-14 μm

In support of system applications which operate in the 3-5 and 8-14 μm spectral ranges, high performance doped silicon detectors are desired to allow for the integration of detectors and silicon electronics on the same substrate. For the 8-14 μm range, detectors were prepared using aluminium, gallium, bismuth and magnesium as dopants which operate in the liquid neon temperature (27 K) range. For the 3-5 μm range, detectors were prepared using indium, sulphur and thallium as dopants which operate in the liquid nitrogen temperature (78K) range. The spectral and temperature characteristics of these detectors are presented. Background limited performance ($\sim 30^\circ$ FOV) is demonstrated for Si:Al, Si:Ga, and Si:Bi up to temperatures of about $30^\circ K$. Background limited performance ($\sim 30^\circ$ FOV) is demonstrated for Si:In and Si:S up to temperatures of approximately 60 and $75^\circ K$ respectively. The data gives a good fit with theory based on background and temperature limitations. The performance of the Si:Mg detector is limited by a shallow unidentified energy level while that of the Si:Tl detector is limited, at present, by low responsivity.

Reprinted with permission from Infrared Physics, Vol. 16, pp 435. Copyright (1976), Pergamom Press.

Sclar, N.

SURVEY OF DOPANTS IN SILICON FOR 2-2.7 AND 3-5 μm INFRARED DETECTOR APPLICATION

In support of the continuing quest of combining i.r. detectors and silicon electronics on a single substrate, an experimental survey was made of doping candidates for use in the 2-2.7 and 3-5 μm i.r. spectral ranges. The elements beryllium, copper, zinc, nickel, thulium and ytterbium were explored for applicability. Measured spectral response curves and the dependence of detectivity on temperature are presented and compared, where applicable, with theory. Various useful attributes of some of these dopants (Si:Cu, Si:Zn) including higher operational temperatures as well as the observed limitations are discussed and critiqued against the Si:In, Si:S and Si:Tl detectors previously investigated for the 3-5 μm range.

Reprinted with permission from Infrared Physics, Vol. 17, pp 71. Copyright (1979), Pergamom Press.

Shin, S. H., Vanderwyck, A. H. B., Kim, J. C. and Cheung, D. T.

HgCdTe PHOTODIODES FORMED BY DOUBLE-LAYER LIQUID PHASE EPITAXIAL
GROWTH

High-performance HgCdTe photodiodes have been formed by successive
growth of p- and n-type epitaxial layers on CdTe substrates via the
liquid phase epitaxy technique. These diodes exhibit high resistance-
area (R_oA) products at high temperatures: R_oA products of 1 and 30
Ω cm^2 have been observed at 283 and 200 K, respectively, for
$Hg_{0.68}Cd_{0.32}Te$ (λ_{co} = 4.0 μm at 200 K). The saturation current
density for the grown junction photodiode at 300 K is 0.12 A/cm^2.

Reprinted with permission from Applied Physics Letters, Vol. 37, No.
4, pp 402, (1980).

Shivanandan, K., McNutt, D. P. and Bell, R. J.

SPECTRAL RESPONSE OF n-TYPE InSb IN THE SUBMILLIMETER RANGE

The spectral response of an n-type InSb bolometer and a Golay
detector were compared between about 10 and 100 cm^{-1}. The optical
systems were held nearly constant and major wave number dependent
deviations were removed from the data. The relative responsivities
are given with the detector, cathode followers, and preamplifiers
treated as a single component with the detector.

Reprinted with permission from Infrared Physics, Vol. 15, pp 27.
Copyright (1975), Pergamom Press.

Sommer, A. H.

THE ELEMENT OF LUCK IN RESEARCH-PHOTOCATHODES 1930 TO 1980

Six photocathode materials were developed during the period from
1930 to 1963 to provide the spectral response and other
characteristics needed for such applications as photometry,
television, scintillation counters, and night vision devices. The
history and the essential properties of these materials are reviewed
and it is shown that all the cathodes resulted from lucky accidents
and not from the application of scientific insight. The period of
empirical innovation came to an end in the late 1960's when negative
electron affinity (NEA) materials became the first photocathodes that
were developed on a strictly scientific basis.

Reprinted with permission from Journal of Vacuum Science and
Technology, Vol. A1, No. 2, pp 119, (1983).

Spicer, W. E.

NEGATIVE AFFINITY 3-5 PHOTOCATHODES: THEIR PHYSICS AND TECHNOLOGY

Negative electron affinity (NEA) photocathodes are defined by the relationship between the potential barrier at the surface and the bottom of the conduction band in the bulk of the material. If the bottom of the conduction band lies above the potential barrier at the surface, the device is said to have a negative electron affinity. In practice this condition is obtained by heavy p-doping of the semiconductor (to encourage downward band bending at the surface) and by adding a thin film (several atomic layers) of cesium rich cesium oxide on the clean semiconductor surface. The physics, development, fabrication, and applications of the NEA cathode are reviewed.

The threshold of response of a NEA photocathode is set by the semiconductor bandgap. By alloying to form ternary or quaternary 3-5 compounds (3-5 compounds are formed from elements of the 3rd and 5th columns of the periodic table), the bandgap (and thus the threshold) can be placed at any desired photon energy within certain limits. The most important limit is that at about 1.1 eV which is the lowest limit achieved for NEA cathodes. This limit is set by the point at which the bandgap of the 3-5 material becomes less than the surface potential barrier. Fundamental work aimed at understanding the 3-5: cesium oxide "interfacial" barrier which sets this limitation is briefly discussed. Because of the "interfacial" barrier, the quantum yield of NEA cathodes decreases as the threshold of response moves to lower photon energy. Field assisted photocathodes provide a means of extending the threshold of response beyond 1.1 eV. Two different approaches to field assisted photocathodes and recent achievements are discussed.

A major advancement has been the achievement of semi-transparent NEA photocathodes by sealing GaAs to glass. This makes possible practical NEA image tubes. The thermionic emission from 3-5 NEA cathodes can be orders of magnitude lower than that form conventional photocathodes. The reasons for this are discussed. Yield and dark current data are given on 3-5 NEA cathodes in operating photomultipliers.

Reprinted with permission from Applied Physics, Vol. 12, pp 115, (1977).

Stevens, N. B.

RADIATION THERMOPILES

Thermocouples and thermopiles have been employed for many years to detect infrared radiation. The thermocouple generates a voltage

due to the Seebeck effect when illuminated with radiation, causing a
temperature rise. The operation and theory of these devices and the
relevant figures of merit to describe their performance are presented.
The character and magnitude of the noise encountered in such radiation
detectors is considered, and finally, typical present day thermopiles,
fabricated from both bulk material and thin films are shown and their
properties described.

Semiconductor and Semimetals, Vol. 5, Infrared Detectors, Chapter 7,
pp 287, (1970). Editors Willardson, R. K. and Beer, A. C., Academic
Press.

Stillman, G. E., Wolfe, C. M. and Dimmock, J. O.

DETECTION AND GENERATION OF FAR INFRARED RADIATION IN HIGH PURITY
EPITAXIAL GaAs

High-purity epitaxial GaAs has been shown to be a fast, sensitive
photoconductive detector in the wavelength range from 100 to 350 μ.
The mechanism of operation involves the photothermal ionization of
shallow donor levels in which bound electrons are first excited from
the ground state to an excited state by the incident radiation and are
then transferred into the conduction band thermally. The spectral
response of the detectors is characterized by a dominant peak at about
282 μ resulting from the excitation of electrons from the ground state
to the first excited state, and by a broad continuum at higher
energies resulting from the photoionization of the shallow donor
levels. The highest detectivity was obtained with material
characterized by a donor concentration of 2.0×10^{14} cm^{-3}, an
acceptor concentration of 4.0×10^{13} cm^{-3}, and an electron mobility at
77° K of 153,000 cm^2/V sec. The NEP of the detector system was
measured using 500° K blackbody source chopped at 260 Hz and filtered
to exclude wavelengths shorter than about 150 μ. The results of these
measurements give an NEP at 282 μ of 5.3×10^{-13} W in a 1 Hz bandwidth.
The variation of the NEP of the GaAs photodectors with donor
concentration is discussed, as well as the effects of lower
temperature and reduced background radiation.

Reprinted with permission from Proc. Symp. Submillimeter Waves, pp
345, (1971). Polytechnic Institute of Brooklyn, New York.

Stillman, G. E., Wolfe, C. M. and Dimmock, J. O.

FAR INFRARED PHOTOCONDUCTIVITY IN HIGH PURITY GaAs

Far infrared photoconductivity in high purity GaAs has been used
to study its shallow donor states and to extend the long wavelength
limit for extrinsic photoconductive detection. The information

concerning the properties of these states obtained from photoconductivity measurements and the performance of GaAs detectors are reviewed. There are also sections on the growth and preparation of GaAs using vapour phase or liquid phase epitaxy, and the materials characterization.

GaAs extrinsic detectors are useful in the spectral range from approximately 100 μm to 400 μm when high sensitivity and speed of response are required.

Semiconductors and Semimetals, Vol. 12, Infrared Detectors II. Chapter 4, pp 169, (1972). Editors, Willardson, R. K. and Beer A. C., Academic Press.

Stokowski, S. E., Venables, J. D., Byer, N. E. and Ensign, T. C.

ION-BEAM MILLED, HIGH-DETECTIVITY PYROELECTRIC DETECTORS

Self-supporting wafers of pyroelectric materials have been prepared in thicknesses as small as 4 μm using ion-beam milling. Normalized detectivities (D*) of 1 mm^2 LiTaO$_3$ detectors fabricated from these very thin wafers have been measured to be 8.5 x 10^8 cm Hz$^{\frac{1}{2}}$ W^{-1} at 30 Hz. It is shown that this technique has great promise in realizing detectors with D* values significantly above 10^9 cm Hz$^{\frac{1}{2}}$ W^{-1}. In addition, the ion milling technique provides the advantages of clean, relatively damage-free surfaces, while permitting the fabrication of detector wafers with complex geometrical structures.

Reprinted with permission from Infrared Physics, Vol. 16, pp 331. Copyright (1976), Pergamon Press.

Stotlar, S. C., McLellan, E. J., Gibbs, A. J. and Webb, J.

10.6 μm DAMAGE THRESHOLD MEASUREMENTS ON SUB-ONE-HUNDRED-ps PYROELECTRIC DETECTORS

Sub-one-hundred-ps response time pyroelectric detectors are being developed at Los Alamos Scientific Laboratory (LASL) to be compatible with the 5-GHz oscilloscope direct-access mode of operation without damage. Strontium barium niobate, lithium tantalate lanthanum-doped lead zirconate, and lithium niobate are being evaluated for use in the edge and complanar configurations. Devices designed at LASL are compared with commercially available detectors. Test results of a less than 15-ps risetime, 31-ps fall time 50/50 SBN pyroelecltric detector are reported. Measurements to date of the damage threshold at 10.6 μm of the above materials in bulk, with various surface treatments, and in devices using 100-ps to 100-μs pulses are also reported.

Reprinted with permission from Ferroelectrics, Vol. 28, pp 325, (1980).

Stupp, E. H.

PYROELECTRIC VIDICON THERMAL IMAGER

Recently a new type of infrared imaging device, the pyroelectric vidicon has been developed, which operates at room temperature. These systems are much cheaper than quantum detectors which generally require cooling. The principals of operation and performance are reviewed. These devices are now being investigated for security sensors, fire location and medical diagnostics. Future systems should have better spatial and temperature resolution using deuterated triglycine fluoroberyllate as target material and reticulation.

Proceedings of SPIE, Vol. 78, pp 23, (1976).

Thomas, C. M.

PHOTOCATHODES

The general properties and principles of photocathode operation are discussed and a brief description of some typical conventional photocathodes, which are commercially available, is given. However, the development of the negative electron affinity devices has enabled high quantum efficiency photocathodes to be produced. GaAs and III-V alloy systems have been fabricated with sensitivities of upto 2000 µA/lumen, and it is expected thyat these NEA photocathodes will satisfy most requirements for image dintensifiers at wavelengths below 1.1 microns. Originally, it was thought they would be useful at wavelengths out to 1.4 to 1.5 microns. However, the internal barrier at the semiconductor/caesium oxide interface precludes this possibility. Field-assisted photocathodes are most likely to be the only method of increasing the response to 2 microns and beyond.

Proceedings SPIE, Vol. 42, pp 71, (1973).

Von Ortenberg, M. and Link, J.

STANNIC OXIDE, A NEW FIR DETECTOR MATERIAL?

We report on the strong bolometric response of doped stannic oxide single crystals to submillimeter radiation. As confirmed up to 120 kG, the magnetic field dependence of the bolometric response is negligible. The bolometer responsivity at zero frequency has been evaluated from the load curve and compared with the corresponding data of the Ge bolometer. The new detector has been used successfully in a magnetic field to detect the electron cyclotron resonance in InSb.

Reprinted with permission from Journal of Optical Society of America, Vol. 67, No. 7, pp 968, (1977).

Wang, C. C., and Lorenzo, J. S.

HIGH-PERFORMANCE, HIGH-DENSITY, PLANAR PbSnTe DETECTOR ARRAYS

Reproducible high-quality photovoltaic detectors have been fabricated by impurity diffusion through well-defined oxide windows into a lead-tin telluride layer grown by liquid phase epitaxy (LPE) on top of a PbSnTe substrate. The carrier concentration of the epilayer was approximately $4 \times 10^{16}/cm^3$ and that of the substrate approximately $5 \times 10^{18}/cm^3$. The oxide layer, used instead of photoresist as an insulator, was found to be pinhole-free and adhered well to the PbSnTe surface. Standard photolithographic and diffusion techniques were used to achieve planar monolithic detector arrays. At 77° K, the average impedance-area product ($R_o A$) was 1.85 Ωcm^2 for an array of 124 2 x 2 mil elements. The average quantum efficiency and peak detectivity (D*) at 11 μm without an antireflection (AR) coating were 40% and $2.6 \times 10^{10} cm$ Hz$^{\frac{1}{2}}$/W, respectively, for a 300° K background and a 180° field of view (FOV).

Reprinted with permission from Infrared Physics, Vol. 17, pp 83. Copyright (1977), Pergamon Press.

Warner, D. J., Pedder, D. J., Moody, I. S. and Burrage, J.

THE PREPARATION AND PERFORMANCE OF RETICULATED TARGETS FOR THE PYROELECTRIC VIDICON

The spatial resolution of the pyroelectric vidicon may be significantly improved by reticulation of the pyroelectric target. This paper describes the development of techniques and structures for the preparation of high quality reticulated vidicon targets in deuterated triglycine sulphate (DTGS).

The dimensions of the reticulated target structures were selected using the results of a theoretical analysis. Reticulated targets were prepared by low energy ion-beam milling of thin targets of DTGS. The ion-beam etching characteristics of this material are described and the selection of masking materials, support layers and the signal electrode are discussed. Large numbers of targets were scanned in vidicon tubes so that differences in imaging performance and cosmetic

appearance could be correlated to changes made to the target
configuration and etching process. From this study an optimum target
preparation and processing route was developed which provided
significant improvement in resolution over tubes possessing planar
targets, and also gave acceptable image appearance.

Reprinted with permission from Ferroelectrics, Vol. 33, pp 249,
(1981).

Watton, R.

PYROELECTRIC MATERIALS: OPERATION AND PERFORMANCE IN THERMAL IMAGING
CAMERA TUBES AND DETECTOR ARRAYS

The present performance of pyroelectric thermal imaging devices is
described. Limiting factors in the operation are then reviewed and
the contribution of pyroelecltric materials research in overcoming
these is discussed.

Reprinted with permission from Ferroelectrics, Vol. 10, pp 91, (1976).

Watton, R., Manning, P., Burgess, D. and Gooding, J.

THE PYROELECTRIC/CCD FOCAL PLANE HYBRID: ANALYSIS AND DESIGN FOR
DIRECT CHARGE INJECTION

Recently attention has been given to the solid state readout of
large pyroelectric arrays, particularly the interfacing of two
dimensional arrays with a silicon charge coupled device (CCD). The
paper gives the theoretical analysis of the electrical and thermal
interface, identifies the critical design parameters and develops the
signal and noise performance for a suitable design. A noise
Equivalent Temperature Difference (NETD) in the region of 0.5^{o}C is
predicted. Experimental measurements on custom CCDs verify the
analysis.

Reprinted with permission from Infrared Physics, Vol. 22, pp 259,
(1982). Copyright © Controller HMSO, London.

Watton, R., Manning, P. and Burgess, D.

INTERFACE DESIGN FOR THE PYROELECTRIC/CCD HYBRID

The use of pyroelectric detectors for inexpensive thermal imaging
systems in particular the vidicon camera tube, operating at ambient
temperatures is now established. To improve the temperature
resolution and the physical characteristics such as size, power

requirements and ruggedisation, attention has now turned to solid state readout of large pyroelectric detector arrays and in particular to the possibility of interfacing the array directly with a silicon CCD chip. A direct injection mode of coupling between the detector and CCD has been chosen and a chip designed to produce a 16 x 16 array. This custom CCD has been manufactured by Plessey Research Limited.

Proceedings SPIE, Vol. 395, pp 78, (1983).

Williams, G. F., Capasso, F. and Tsang, W. T.

THE GRADED BANDGAP MULTILAYER AVALANCHE PHOTODIODE: A NEW LOW-NOISE DETECTOR

We propose a new multistage avalanche photodiode for low-noise optical detection. In each stage, the ionization energy is provided by a heterointerface conduction-band step. Thus, ideally, only electrons cause ionization, and the device mimics a photomultiplier. This detector has intrinsically lower noise and lower operating voltage than conventional avalanche detectors. Designs for 1.3 μm fibre optic systems are presented and possible realizations using molecular beam epitaxy discussed.

Copyright © 1982 IEEE. Reprinted with permission from IEEE Transactions on Electron Device Letters, Vol. EDL-3, No. 3, pp 71, (1982).

Woodhead, A. W.

IMAGE TUBES AND THEIR APPLICATIONS

Three common forms of image tube are discussed, a proximity focussed tube, a magnetically focussed system and an electrostatically focussed tube based upon a system of concentric spheres (cathode and anode). The gain of these systems can be increased by using a cascade tube or a channel multiplier tube. The basic noise mechanisms which limit the performance of an image intensifier are photon noise, dark emission and noise generated by the gain mechanism. The major applications for these devices are as night vision aids, in X-ray imaging and for high speed photography.

Vacuum, Vol. 30, pp 539, (1980).

Yamaka, E., Teranishi, A., Nakamura, K. and Nagashima, T.

PYROELECTRIC VIDICON WITH GROOVED RETINA OF $PbTiO_3$ CERAMICS

To improve the spatial resolution of $PbTiO_3$ pyroelectric vidicons, a surface of $PbTiO_3$ ceramic wafer was grooved by Q-switched YAG laser. Its Curie temperature is so high that no poling process is required after polishing to a thin target and outgassing in evacuation. The effective reduction of a lateral heat conduction due to grooving was observed in time variations of signal outputs to a step radiation in a shutter mode. The experimental reduction ratio seems to support a simple theory that the reduction ratio is equal to a ratio of the target thickness to the residual thickness.

Reprinted with permission from Ferroelectrics, Vol. 11, pp 305, (1976).

Yokozawa, F.

TECHNOLOGY TREND ON THE PHOTOMULTIPLIERS

Photomultiplier tubes are suitable for the detection of radiation from ultraviolet to the near infrared and are sensitive enough to detect a single photon. In general the overall amplification of the dynodes may reach 10^6 to 10^7. The properties of the most popular photocathodes and dynode materials are discussed. The future trends will be aimed at the detection of weaker and faster signals and for two dimensional measurements which might require the use of a micochannel plate dynode, a mesh type dynode, a multianode or a resistive anode which will be able to define the position of an incident photon.

JEE (Japan), Vol. 19, pp 94, (1982).

Zwicker, H. R.

PHOTOEMISSIVE DETECTORS

There are three general detector applications for which photoemissive devices are uniquely suited, these are the detection of low intensity signals, high speed detection of low level signals and the acquisition of high resoslution spatial information. The physical principles of both classical and negative electron affinity photoemissive surfaces are discussed, and the operation and construction of both types of devices are then examined in detail, in particular the CsSb and AgCsO classical and the GaAs NEA photocathodes. Other infrared sensitive emitters, including the quaternary alloy systems such as InGaAsP are then briefly summarized. The chapter is concluded with a summary of device-to-device trade-offs in classical and NEA devices.

Topics in Applied Physics, Vol. 19, Optical and Infrared Detectors, pp 149, (1977), Editor Keyes, R. J., Springer-Verlag.

INDEX

173